Weather and Climate Science Education for Grades 3–12

Weather and Climate Science Education for Grades 3–12

Natalie J. Carroll,
Theodore Leuenberger
Gretchen Leuenberger

Purdue University Press
West Lafayette, Indiana

 Cataloging-in-Publication Data on file at the Library Congress.

978-1-62671-405-2 (paperback)
978-1-62671-406-9 (epdf)

Cover image courtesy of Natalie Powell.

Any figures available in color appear that way in the e-book format.

Contributors: Purdue University; Natasha Duncan, Olivia Kellner, and Jesse Steinweg-Woods (former students); Sam Lashley, Evan Bentley, Courtney Obergfell, Amos Dodsonand, and Peggy Keller (reviewers); Aime Lillard and Sayde Uerkwitz (administrative assistants); Carol Bloom, Nancy Alexander, and Neal Novak (copyediting); Natalie Powell (Just Natalie Graphic Design, design and layout).

Contents

Level 1

You should be able to complete each activity in about 20–30 minutes. Some activities may take a shorter time. Some may also take longer, particularly when completing a Fly Higher activity.

Level 2

Level 3

Weather

Climate

Facilitator's Guide

Activity Facilitation and Answers: Level 1 225

Climate

Science Standards

LEVEL 1

Weather and Climate Science Education for Grades 3–5

Note to the 4-H Member

Everyone seems interested in the weather. We need to know what the weather will be so we can dress appropriately. That helps us plan our day. Gardeners and farmers need to understand the area's climate. That helps them know what they can plant and what will grow. Homeowners also need to understand our climate. It helps them know which flowers and shrubs to plant around their houses. It also tells them what kind of furnace and air conditioner will work best in this area. And when stormy weather is predicted, people need to be prepared. It is important that they know how to be safe in bad weather, whether they are in their homes or outside.

LEARNING ABOUT WEATHER AND CLIMATE

The Level 1 of *Weather and Climate Science for Grades 3–12* introduces basic weather words and ideas. These activities focus on understanding the signs of weather around you. You will also begin to learn the difference between weather and climate and how these ideas are related.

When you are in grades 6–8, you will use the Level 2. It will introduce you to more complex weather topics. You will learn about air pressure, winds, humidity, and fronts.

When you are in high school, you will use the Level 3. It tells you even more about weather and climate science. This will help you prepare to fully understand weather all through your lifetime.

The two keys to learning in this—or any—project are to enjoy your studies and to learn at your own pace. We hope this project is just the start of a lifetime understanding the climate you live in and watching the weather!

WHAT YOU'LL LEARN

Weather is the state of the atmosphere at a particular place and time. The climate of the area and many local factors influence the weather.

Climate is the usual, long-term weather conditions for an area or for the entire planet.

Climate system is a complex process in which the earth's water and gas flow or change state because of the sun's energy.

KNOWING THE DIFFERENCE

Weather—The current weather condition of the atmosphere, including temperature, wind, cloudiness, precipitation, and relative humidity.

Climate—The average weather over time.

COMPARING CLIMATES

How are four U.S. climate areas—temperate, desert, tropical, and tundra—the same, and how are they different?

intro Weather is the current condition of the **atmosphere**, including temperature, wind, cloudiness, **precipitation**, and **relative humidity**. Climate is the average weather over time. The local climate affects the clothes you wear, when we plant flowers and vegetables, and the crops that farmers grow. There are four climates in the United States: temperate, desert, tropical, and tundra.

A *temperate* climate has seasons with hot and cold weather, but not as hot or cold as in other climates. Indiana has a temperate climate. We also have rain and snow but, generally, not as much as in some other climates. The *desert* climate has very little rainfall and hot temperatures. *Tropical* climates have warm or hot conditions year-round. The *tundra* climate has extremely cold conditions most of the year. All of these climates have seasons, but the seasons are not always as noticeably different as the seasons we experience in Indiana. You will compare data from states in the four U.S. climate areas—temperate, desert, tropical, and tundra—and look for what is similar and what is different.

OBJECTIVE: Compare the four climates found in the United States.

Weather Tote

- Pencil
- Dictionary

do it

1 Read the data in the chart below for each of the four climates.

2 Answer the questions about the four climates on the next page.

3 Answer the follow-up questions.

	January (winter)	July (summer)	yearly average
TEMPERATE CLIMATE Central Indiana			
average temperature	25.5°F	75.4°F	52.3°F
average precipitation	2.3″	4.5″	39.9″
DESERT CLIMATE			
average temperature	53.6°F	93.5°F	72.6°F
average precipitation	0.7″	0.8″	7.7″
TROPICAL CLIMATE Southern Florida			
average temperature	67.2°F	82.6°F	75.9°F
average precipitation	2.0″	5.7″	55.9″
TUNDRA CLIMATE			
average temperature	-10.1°F	62.5°F	26.9°F
average precipitation	0.5″	1.9″	10.9″

Graphic Illustrations by Jessica Harsh

WHICH CLIMATE HAS:

temperate?
desert?
tropical?
tundra?

The highest average temperature? ____________________ What is it? __________

The lowest average temperature ____________________ What is it? __________

The highest July temperature? ____________________ What is it? __________

The lowest July temperature? ____________________ What is it? __________

The highest January temperature? ____________________ What is it? __________

The lowest January temperature? ____________________ What is it? __________

The highest average precipitation? ____________________ What is it? __________

The lowest average precipitation? ____________________ What is it? __________

The highest July precipitation? ____________________ What is it? __________

The lowest July precipitation? ____________________ What is it? __________

The highest January precipitation? ____________________ What is it? __________

The lowest January precipitation? ____________________ What is it? __________

Look up the word *temperate* in the dictionary, and write the definition below:

____

____

____

____

The first question in the chart above asked you about climate extremes—highest and lowest of the averages. How many times did the temperate climate have the extreme data for one of the climates?

____

SHARE WHAT HAPPENED:

- Were you surprised by the differences in the climates?
- What data surprised you?

APPLY:

- What do you think it would be like to live in each of the other climates?
- Which climate do you think that you would be happiest living in year-round? Why?
- How are flowers, plants, and agricultural crops affected by climate?

GENERALIZE TO YOUR LIFE:

- Why doesn't Indiana have many extremes in temperature and rainfall?
- Temperature and rainfall are the major factors in a climate. What might some other factors be?

Fly Higher

- Calculate the difference in average temperature and average precipitation between January and July for each climate. What climate has the largest extremes for temperature? For precipitation?
- Keep track of temperature and rainfall in several different areas around the United States for a week or for one day during the first week of the months of January, April, July, and October.
- Explore what the climate is in five places around the world where you would like to go someday.

Notes

COUNTRY OF COLORS

What is a hardiness zone?

Buds, new leaves on trees, tulips, and daffodils tell us that spring is coming because the weather is warming up. These plants can survive our cold winter temperatures. But many flowers we enjoy need to be planted in the spring after the danger of frost has passed. In this activity, you will learn about **hardiness zones** and how they are determined.

The U.S. Department of Agriculture (USDA) publishes the Hardiness Zone Map to help gardeners and farmers plan what and when to plant. The map shows zones based on the **average annual extreme minimum temperatures** for a 30-year period in the United States. This is important information because many plants are very sensitive to cold temperatures and cannot survive freezing.

AVERAGE ANNUAL EXTREME MINIMUM TEMPERATURES

Average *the sum of all values divided by the number of years*

Annual *over a year*

Extreme *the lowest*

To find the average annual extreme minimum temperature, add the lowest temperature of each year for 30 years and divide the sum by 30. Areas with similar extreme minimum temperatures (within 10°F) are grouped together in zones. Areas with colder winter weather have lower zone numbers. Areas with warmer winters have higher zone numbers.

OBJECTIVE: Know Indiana's hardiness zones.

Question: Why do some flowers bloom year after year, while others need to be planted each year?

____

____

____

Weather Tote

- Pencil
- Colored pencils, markers, or crayons
- Hardiness Zone Map (on next page)

1. Draw a line between Indiana's hardiness zones.
2. Color each zone a different color.
3. Make a key that shows the color for each zone.
4. Place an "x" where you live.
5. List five garden plants (flowers or vegetables) that cannot overwinter—not survive a winter and need to be planted each year.

PLANT HARDINESS ZONES

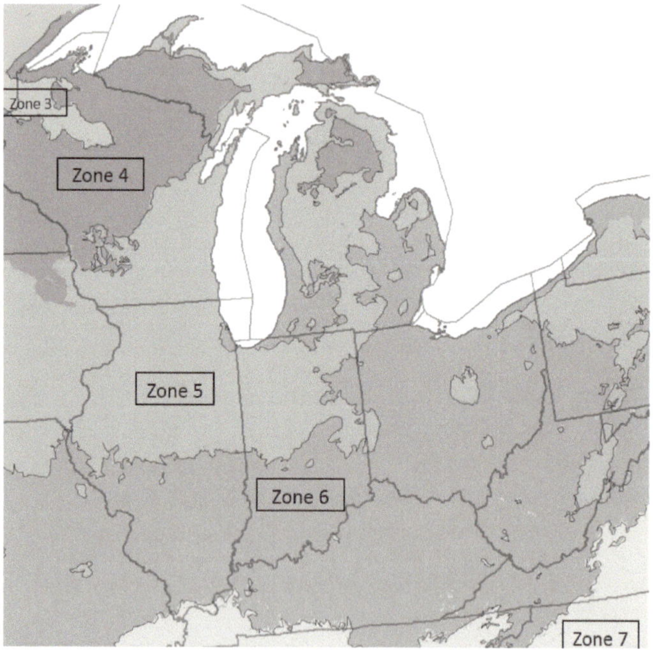

Adapted from the USDA Hardiness Zone map, planthardiness.ars.usda.gov

INDIANA HARDINESS ZONES

Garden plants that cannot overwinter

1 ______________________________

2 ______________________________

3 ______________________________

4 ______________________________

5 ______________________________

SHARE WHAT HAPPENED:

How many zones did you find in Indiana?

APPLY:

- What is the average annual extreme minimum temperature range where you live?
- What do you think would happen if you planted a hardy perennial flower, a plant that lives more than one year, in Hardiness Zone 9, which includes the warmest parts of the United States.

GENERALIZE TO YOUR LIFE: How do the hardiness zones affect what farmers grow?

Fly Higher

- Visit the USDA Plant Hardiness Zone Map online (and in color) at http://planthardiness.ars.usda.gov/PHZMWeb/Maps.aspx, or look at the map to the left. Then answer the following questions.

  Where are the warmest zones? What is their average minimum temperature?

  Where are the coldest zones? What is their average minimum temperature?

- Find a flower or vegetable seed catalog. Your parent or Extension Educator can help you.

  How are the hardiness zones listed for different flowers or vegetables?

  What flowers or vegetables did you find that are not recommended for where you live?

DEFINING WEATHER WORDS

What words do I need to know to study and understand the weather?

intro Do you know what the four H's stand for in "4-H?" Sometimes people use words that others don't know. Do you have a favorite sport that uses words some of your friends don't know? Scientists who study weather and climate use both common words and special terms that others may not know. This activity will help you get started learning weather words.

OBJECTIVE: Learn common weather words.

Weather Tote

- Pencil

do it

1 Rearrange the scrambled letters on the next page to make weather words. Use the Word Bank *only* if you cannot figure out the words. Then describe each weather word in your own words.

Chat

SHARE WHAT HAPPENED:

- Did you use the Word Bank, or did you do this activity without looking at the Word Bank?
- Were some words harder to unscramble than others?
- Were some words harder to define than others?

APPLY: How does knowing weather words help you talk about the weather?

GENERALIZE TO YOUR LIFE:

- Describe how you dress when it is raining.
- How do you dress when it is snowing?
- How do you dress when it is hot outside?

Fly Higher

- Compare your weather definitions to the ones you find in a dictionary.

Scrambled Letters	Weather Word	My Definition
ahrewet		
rai		
tarew		
athe		
miudtihy		
nsu		
herta		
nwid		
nswo		
nria		
lhia		
scoldu		
ateuprrmeet		
etamilc		

Word Bank

- air
- clouds
- hail
- rain
- sun
- weather
- humidity
- climate
- earth
- heat
- snow
- water
- wind
- temperature

Notes

EARTH'S SURFACES

How will the energy from sunlight heat up different-colored surfaces?

Sunlight warms the surfaces of earth. Look around you. Do all of earth's surfaces look the same? Earth has many natural surfaces, including grass, trees, sand, water, mountains, and ice. Humans create many surfaces, including roofs, sidewalks, and roads of concrete or asphalt.

When surfaces are warmed by sunlight, they warm the air above them. Different surfaces may warm the air at different rates or to a higher or lower degree. Surface warming can affect our weather.

OBJECTIVE: Study how the color of a surface influences how fast an object warms up.

Weather Tote

- Indoor thermometer, with a bulb that can safely touch the paper
- Desk lamp with a 40-watt or 60-watt bulb
- Clock or watch with a second hand
- One sheet each of white, black, brown, green, and blue construction paper

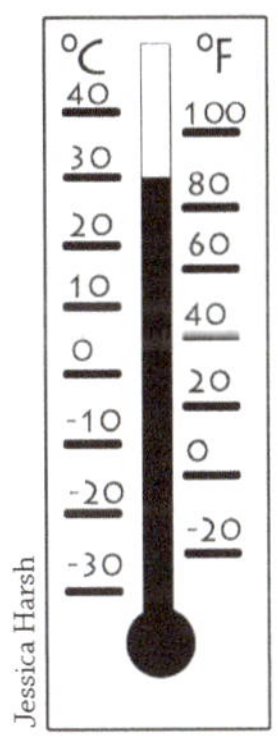

1. Record the room's temperature on the thermometer as the beginning temperature in the chart on the next page.
2. Place the white sheet of paper on top of the thermometer.
3. Put the desk lamp 6 inches above the paper at the bulb end of the thermometer.
4. Turn on the light and wait 5 minutes.
5. Record the temperature on the thermometer as the ending temperature on the chart.
6. Calculate the change in temperature by subtracting the first temperature from the second one. Record it on the chart.
7. Allow the thermometer to cool for 5 minutes.
8. Repeat the steps for the other colors of paper, and fill in your results in the chart.

SHARE WHAT HAPPENED:

- Which color did you think would heat the most?
- Which color did heat the most?
- Which color heated the least?
- What surfaces on earth would match the colors used in your experiment?
- Which surfaces would heat the air above them the most? The least?

APPLY: Why does the sun's uneven heating of oceans and land surfaces cause breezes?

Hint: Have you ever been to a sandy beach on a hot, sunny day? What happens if you go barefoot on the sand? How does it feel when you sink your feet in the water? Both surfaces—sand and water—receive the same amount of sunlight, but each heats up much differently.

Look at the arrows in the diagram in column 2 that show a sea breeze. Then answer the questions.

- Why is the air over the land rising?
- What is causing the sea breeze?
- How does the surface affect breezes?

SUN'S HEAT AND SEA BREEZES

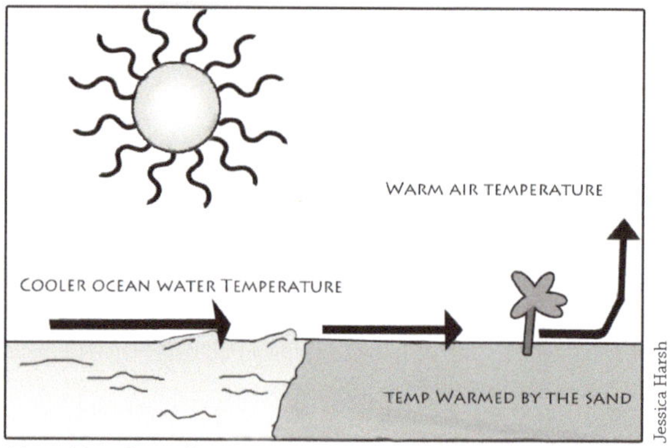

Adapted from https://www.noaa.gov/jetstream/ocean/sea-breeze

GENERALIZE TO YOUR LIFE: Think of places that have two different surfaces, like a plowed field next to a forest or a large parking lot next to a city park.

- How might the different surfaces affect the wind in each place?
- How might humans change the earth's surfaces and affect the local weather?

Fly Higher

Take an outdoor thermometer outside on a sunny, calm day. Put the thermometer on several different surfaces to see how the sunlight is causing these surfaces to warm. Use something to shade the thermometer (like an umbrella) so that the sunlight does not affect the thermometer.

TABLE 8. TESTING COLORED SURFACES

Surface Color	Beginning Temperature	Ending Temperature	Temperature Change
White			
Black			
Brown			
Green			
Blue			

H_2O

What causes water to condense out of the air?

Water is important for all life on earth. It plays a big part in determining the weather. Water has three forms—solid, liquid, and gas. Water on earth's surfaces **evaporates** into the air as a gas, forms clouds in the sky, and creates **precipitation**. It might be snow, rain, or **water vapor**. When water falls from the sky as rain, it can soak into the ground. It can run along the surface to form streams and rivers. It can collect in areas and form ponds and lakes. It can travel to the ocean. Plants, people, and other animals also use water. During all of these possible journeys, the water can evaporate back into the air, and what we call the water cycle starts all over again.

OBJECTIVE: Recognize how water forms precipitation.

Weather Tote

- Two glass jars (about pint sized) with lids
- Water
- Ice cubes

PART 1

1 Fill a jar with water at about room temperature, and put on the lid.

2 Fill the second jar with ice cubes. Then completely fill the jar with cold water. Put the lid back on.

3 Observe the two jars for about 5 minutes.

SHARE WHAT HAPPENED:

- On which jar did water **condense** on the outside?
- Where do you think the condensed water came from?
- Why did water only condense on the jar with the ice cubes?

PART 2

1 Take the first jar, the one without ice, and empty most of the water. Leave about a centimeter (½ inch) of water in the bottom of the jar.

2 Put on the lid, and shake the jar for a full minute. This will force water to **evaporate** in the jar.

3 Set the jar down, and leave it alone for another minute.

4 Observe what the jar looks like.

5 With the lid on tight, hold the first jar sideways over a sink.

6 Carefully pour the ice water from the second jar down the side of the first jar.

7 Observe the inside of the first jar.

SHARE WHAT HAPPENED:

- What did you observe on the inside of the jar at each step?
- What happens to the temperature of the air when water condenses?

APPLY: Study the picture of the water cycle. How does what you learned about precipitation relate to the water cycle?

GENERALIZE TO YOUR LIFE:

- When does dew form? When does frost form? Would you ever expect dew or frost to form when a surface is warming up?
- Have you ever watched clouds on a sunny day? What do they look like?

THE WATER CYCLE https://www.noaa.gov/education/resource-collections/freshwater/water-cycle

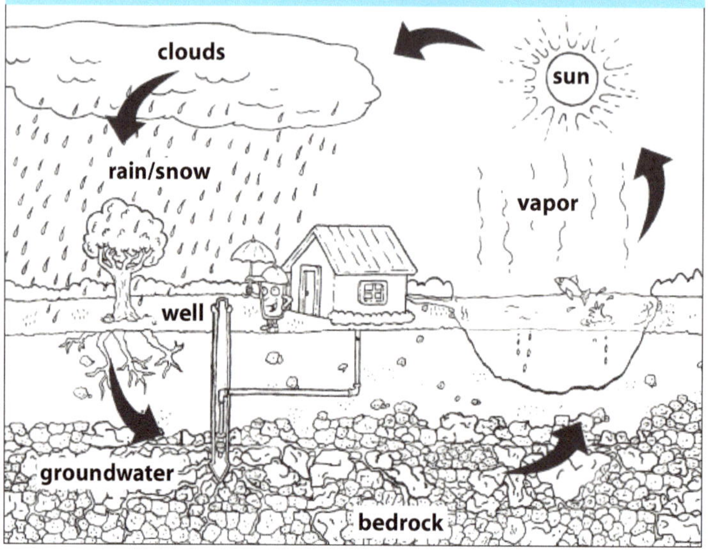

INVISIBLE AIR

How do we know air is there?

Air is one of many things we know exists but that we cannot see. Earth is covered with a layer of air called the **atmosphere**. Weather occurs in the atmosphere. When the air interacts with the sun's energy and earth's surface, it creates weather. So you need to understand the properties of air to understand weather.

OBJECTIVE: Understand that warm air rises and cool air sinks.

PART 1

Question: Does air take up space?

Weather Tote

- Glass
- Clear bowl
- Water
- Towel
- Ruler

1 Place the clear bowl on the towel.

2 Add water so the bowl is about ¾ full.

3 Measure the height of the water:

____

4 Turn the glass upside down, and push it straight down into the water.

5 Measure the height of the water again:

____

6 Record the difference in the height of the water before (3) and after (5):

____

SHARE WHAT HAPPENED:

- What happened to the water in the bowl when you inserted the upside-down glass?
- What kept the water from entering the glass?

PART 2

Question: How does temperature change the amount of space that air occupies?

Weather Tote

- Glass jar (not plastic) with a metal screw-on lid (pint or similar size will work)
- Clear skinny plastic straw, like some restaurants use
- Hot glue gun and glue stick or Play-Doh
- Food coloring
- Ice cubes or freezer cold packs
- Towel
- **Also Needed:** Ask an adult to drill a hole in the center of the metal lid. Using a drill bit the same size as the diameter of the straw, he or she should place the metal lid on the block of wood, hold the lid and drill a hole in the center.

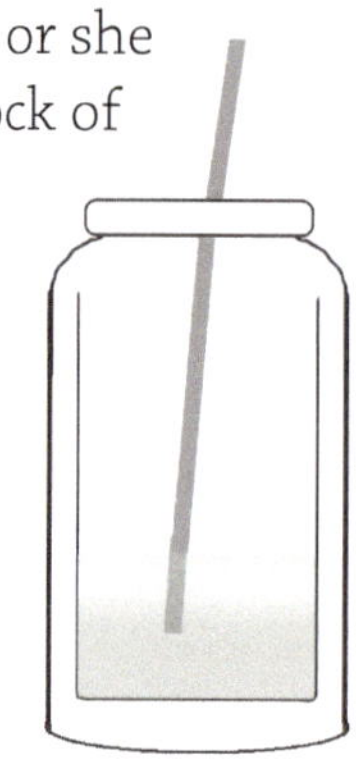

What do you think will happen when air is cooled?

To help you answer this question, construct a testing jar using the following steps:

1 Place the lid on the jar, and put the straw through the hole so that the bottom of the straw is about ½" from the bottom of the jar.

2 Seal the straw to the top of the lid, using hot glue or Play-Doh.

3 Remove the lid from the jar, and seal the straw to the bottom of the lid for an airtight seal.

4 Place about 1" of water in the jar, and add one drop of food coloring.

5 Place the lid with the straw firmly on the jar.

6 Put the jar on the towel to avoid a mess from spilling water.

7 Rub two ice cubes or freezer cold packs against the outside of the jar to cool the air inside the jar.

8 Observe what happens in the straw as the air inside the jar cools.

9 Now warm the air in the jar by wrapping your hands around the jar.

10 Observe what happens in the straw as the air inside the jar warms up.

SHARE WHAT HAPPENED

- What did you observe when you cooled the air in the jar with the straw?
- What did you observe when you warmed the air in the jar with the straw?

PART 3

Questions:

1 Which is heavier (denser)—warm or cold air? Circle your guess:

WARM COLD

2 If the air in a balloon is cooled, how will the balloon's weight (**density**) change? Circle your guess.

BECOMES HEAVIER (denser)

BECOMES LIGHTER (less **dense**)

NO CHANGE

Weather Tote

- Yardstick or similar board
- Two rubber bands
- Two balloons
- Ice cubes or freezer cold packs
- Masking tape
- Towel

Do it

1. Blow up the balloons, and attach them to each end of the yardstick with the rubber bands.
2. Balance the yardstick on the back of a chair. Tape the balancing point of the yardstick to the chair back.
3. Rub ice cubes on one of the balloons, or put one of the balloons in the freezer or a bowl of ice water for five minutes. Put it back in the same place on the yardstick. If you use ice, avoid a mess by keeping the towel under the melting ice.
4. Observe how the balance changes.

Chat

SHARE WHAT HAPPENED: Did the balloons on the yardstick stay balanced when you cooled one of the balloons?

APPLY: Why does cooled air sink? Why does warmed air rise?

GENERALIZE TO YOUR LIFE:

- Is there a breeze outside today? How might changes in the air temperature have caused that breeze?
- Have you ever seen a hot air balloon? How do you think a hot air balloon works?

Fly Higher

- Using an indoor thermometer, compare the temperature of the air in a room near the floor and near the ceiling. If your house has two or more floors, compare the floor and ceiling temperatures in an upstairs room.
- If you have a garage or barn, compare the two temperatures in that building on a warm day.
- How can you explain any differences?

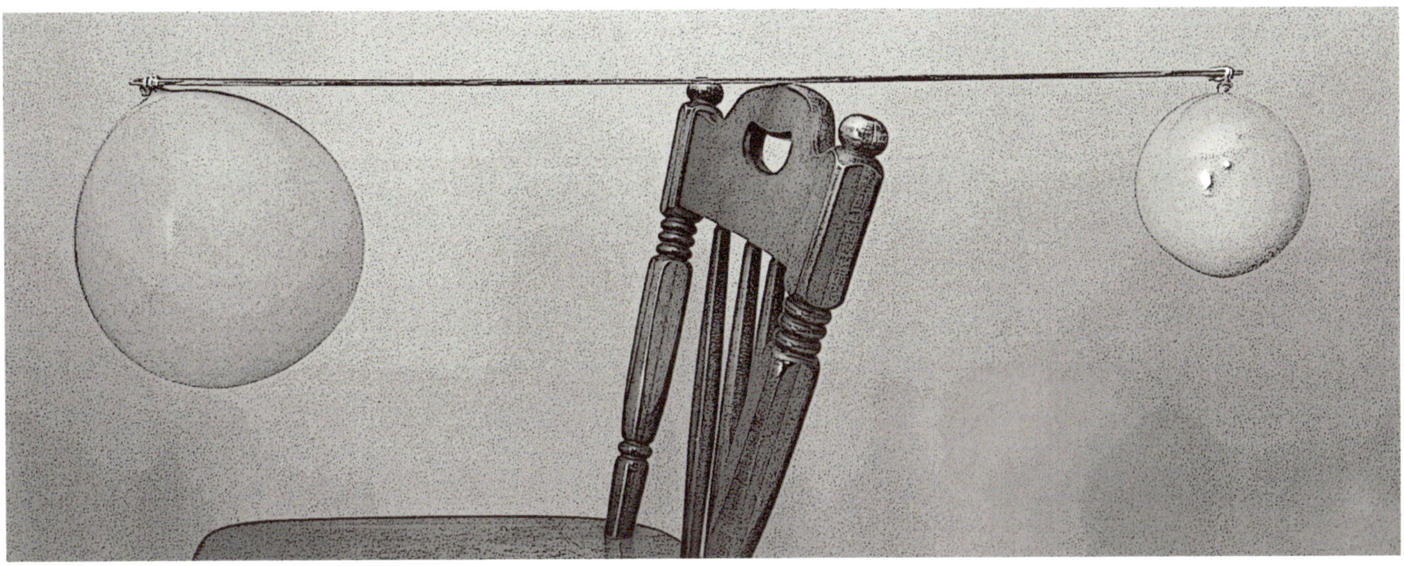

READING ABOUT WILD WEATHER

What damage can severe weather cause?

intro The Wild Weather series by Scholastic includes books on tornadoes, blizzards, floods, and lightning. In this activity, you will choose one of these books to read and learn more about weather. These books can be found in libraries or from an online book source. The following information will help you find these books.

- Hopping, Lorraine Jean, and Jody Wheeler. *Wild Weather: Tornadoes!* New York: Scholastic, 1994. Print. ISBN: 0-590-46338-1
- Hopping, Lorraine Jean, and Jody Wheeler. *Wild Weather: Blizzards!* New York: Scholastic, 1998. Print. ISBN: 0-590-39730-3
- Hopping, Lorraine Jean, and Jody Wheeler. *Wild Weather: Floods!* New York: Scholastic, 2000. Print. ISBN: 0-439-08757-0
- Hopping, Lorraine Jean, and Jody Wheeler. *Wild Weather: Lightning!* New York: Scholastic, 1999. Print. ISBN: 0-590-52285-X

OBJECTIVE: Read a book about severe weather events, become familiar with that type of severe weather event, and describe how to prepare for severe weather.

Weather Tote

- One book from Scholastic's Wild Weather series

Do it

- Read the book you have chosen.
- Report one thing you learned in each chapter:

Chapter 1: ______________________

____

____

Chapter 2: ______________________

____

____

Chapter 3: ______________________

____

____

Chapter 4: ______________________

____

____

Chapter 5: ______________________

____

____

- Discuss the Chat questions.

SHARE WHAT HAPPENED:

- What was the most interesting fact you learned in each of the five chapters?
- Each Wild Weather book introduces a scientist who studies a weather event. Who was the scientist you read about, and what kind of work does he or she do?

APPLY: What kind of conditions can cause the weather event you read about? How can you protect yourself during this type of weather event?

GENERALIZE TO YOUR LIFE:

- Have you had a personal experience with the wild weather event you read about?
- What have you read or heard from news reports about this type of weather event?

Fly Higher

- Write your own wild weather story about an experience you and your family or friends have had.
- Read other books in the Wild Weather series.

NOTES

’TIS THE SEASON

How do the changing seasons affect the way we live?

intro The weather in Indiana changes with the four seasons. These changes cause people to wear different clothes and do different outdoor activities. The seasons also affect the plants and animals that live outside.

OBJECTIVE: Identify the effects of changing seasons.

do it Show how seasons are different by drawing in the boxes what you or others like to see, hear, or do during each season. You might show the clothes people wear outside or the outdoor sports and activities they play, the view outside your window, or how plants change.

SPRING

SUMMER

FALL

WINTER

SHARE WHAT HAPPENED:

- Was it hard to decide how to draw the seasons?
- Was one season more difficult to show than others?

APPLY: What is your favorite season? Why?

GENERALIZE TO YOUR LIFE:

- How does the season affect what you wear?
- How does the season affect what you do outside?
- How does the season affect plants?

Fly Higher

Take photographs, make a video, or find pictures in magazines or on the Internet that show the different seasons. Use your pictures or photos to teach younger children about the seasons.

WATCHING THE WIND

Question: How can you estimate wind speed?

intro More than 200 years ago, in 1806, Sir Francis Beaufort of Great Britain's Royal Navy developed a method for estimating wind speed by making observations. His system is known as the Beaufort Wind Scale. Because he was part of the navy, Sir Beaufort observed the surface of the sea. Over the years, scientists have adapted and modernized his method to produce a very reliable way to estimate the wind speed just by watching the weather outside a window.

OBJECTIVE: Estimate the current wind speed by using the Beaufort Wind Scale.

Weather Tote

- Beaufort Wind Speed Scale (Table 1)
- Radio, TV with Weather Channel access, or Internet—visit https://www.noaa.gov

Do it **NOTE:** Many youth will only be able to observe land conditions to determine the wind speed. You may not be able to make sea observations like Sir Beaufort, unless you live (or vacation) near the ocean or a large lake.

1 Record the date and time in the first two columns of the Wind Speed Data table (Table 2).

2 Study the Beaufort Wind Speed Scale (Table 1) to determine which conditions most closely resemble the wind conditions you see at your home.

3 Enter the wind speed (column 3) in your Wind Speed Data table.

4 Check your weather source immediately after recording your data, and enter the reported wind speed in column 4 of the Wind Speed Data table.

5 Take one to three readings each day.

6 Allow at least 4 hours between readings.

7 Take a total of 15 to 20 wind-speed readings before discussing the questions with your teacher.

Chat **SHARE WHAT HAPPENED:**

- Was it difficult to use the Beaufort Wind Scale?
- How did your readings compare to the actual wind speeds reported from your weather source?
- Do you think the Beaufort Wind Scale is useful?

APPLY: What are some other objects that are affected by the wind that could help you in estimating wind speed?

TABLE 1. BEAUFORT WIND SPEED SCALE

Beaufort Number	Wind Description	Wind Speed	Water and Sea Surfaces	Land Conditions
0	Calm	0-1 mph*	Smooth, mirrorlike	Calm • Smoke rises vertically
1	Light Air	2-3 mph	Ripples without crests	Smoke drift indicates wind direction • Leaves and wind vanes are stationary
2	Light Breeze	4-7 mph	Small wavelets • Crests of glassy appearance, not breaking	Wind felt on exposed skin • Leaves rustle • Wind vanes begin to move
3	Gentle Breeze	8-12 mph	Large wavelets • Crests begin to break, scattered whitecaps	Leaves and small twigs constantly moving, light flags extended
4	Moderate Breeze	13-17 mph	Small waves with breaking crests • Fairly frequent whitecaps	Dust, leaves and loose paper lifted • Small branches begin to move
5	Fresh Breeze	18-24 mph	Moderate waves of some length • Many whitecaps • Small amounts of spray	Branches of a moderate size move • Small trees begin to sway
6	Strong Breeze	25-30 mph	Long waves begin to form • White foam crests are frequent • Some airborne spray	Large branches in motion • Whistling heard in overhead wires • Umbrella use difficult • Empty plastic bins tip
7	Near Gale	31-38 mph	Some foam from waves is blown into streaks • Moderate amounts of airborne spray	Whole trees in motion • Effort needed to walk against wind
8	Gale	39-46 mph	Moderately high waves with breaking crests • Well-marked streaks of foam • Considerable airborne spray	Twigs broken from trees • Cars driving on roads affected • Walking is difficult
9	Strong Gale	47-54 mph	High waves with crests rolling over • Dense foam blowing • Large amounts of airborne spray reduce visibility	Some branches break off trees, and some small trees blow over • Construction and other temporary signs and barricades blow over
10	Storm	55-63 mph	Very high waves and large patches of foam • Considerable tumbling of waves • Large amounts of airborne spray reduce visibility	Trees are broken off or uprooted • Structural damage is possible
11	Violent Storm	64-73 mph	Exceptionally high waves • Very large patches of foam • Very large amounts of airborne spray severely reduce visibility	Widespread vegetation and structural damage is likely
12	Hurricane Force	>73 mph	Huge waves • Sea is completely white with foam and spray • Air is filled with driving spray, greatly reducing visibility	Severe widespread damage to vegetation and structures • Debris and unsecured objects are blown about

*mph = miles per hour

GENERALIZE TO YOUR LIFE: When would it be useful for you to estimate the wind speed? Are there particular places or situations where this knowledge might come in handy?

Table 2 is on the next page.

Fly Higher

Develop your own scale for estimating wind speed by observing a flag on a pole. You might also experiment with different sized flags or pieces of fabric or by placing the flags (or fabric) at different heights.

TABLE 2. **WIND SPEED DATA**

Date	Time	Beaufort Scale Wind Speed	Actual Wind Speed

WEATHER AFFECTS PLANS

How does the weather affect your plans?

It is helpful to know the weather forecast before you leave home every day. If rain is expected, you may wish to wear a raincoat or take an umbrella. If it is going to be very hot, you might take a bottle of cold water. If it is cold, you'll probably wear a warm coat with a hat, boots, and mittens or gloves. In this activity, you will collect weather data and list the type of outdoor activities you can do.

OBJECTIVE: Collect weather data.

Weather Tote

- Your choice for the local weather report—radio, TV, or Internet
- Weekly Weather Chart (Table 3)

1 Use your weather source to obtain the local high and low temperature each day for one week.

Recommended website:
NOAA, www.noaa.gov
Enter your zip code or your city's name and state abbreviation in the Search window in the left-hand sidebar on the home page.

2 Record the temperatures in columns 2 and 3 in the Weekly Weather Chart (Table 3).

3 Indicate if you had any precipitation at your home (columns 4-6).

4 List any outdoor activities that you did each day (column 7).

SHARE WHAT HAPPENED:

- What was the highest temperature during the week you collected data?
- What was the lowest temperature during the week you collected data?

APPLY:

- Did the weather affect your outdoor activities? Why or why not?
- When is the weather most likely to affect your outdoor activities?
- When has the weather affected your family's plans?

TABLE 3. **WEEKLY WEATHER CHART**

	Temperature		Precipitation			Outdoor Activities
	High	Low	Rain	Snow	Other	
Sunday						
Monday						
Tuesday						
Wednesday						
Thursday						
Friday						
Saturday						

GENERALIZE TO YOUR LIFE: What effect does weather have on vacation plans? On building a house? On walking a trail or hiking?

Fly Higher

- Collect weather data for a week during a different season. How do your results compare with the data in Table 1?
- List the outdoor activities that you like to do during Indiana's four seasons: winter, spring, summer, and fall.
- Visit the NOAA website, www.noaa.gov, and enter your zip code (or city and state) on the Weather.gov Forecast search box on the left-hand side. Click Go.

  Read the forecast.

  Scroll down to the hourly weather graph on the lower right-hand side of the forecast page.

  Click on the graph.

  Use the five graphs that appear to review the weather predictions for the next 24 hours, including:

  1 Heat index/dew point/temperature

  2 Wind/gusts

  3 Relative humidity/precipitation potential/sky cover

  4 Rain (**NOTE:** The bars show the likelihood of rain; the totals are given for various time spans across the bottom of the graph.)

  5 Thunder
- Use the data shown to summarize the detailed weather forecast for the next 24 hours for your students, your parents, or your friends.

WEATHER ALERTS

What are NWS alerts?

intro The National Weather Service (NWS) watches the weather for us. When bad weather is approaching, the NWS sends alerts to radio, TV, and Internet sources so that we can prepare to protect ourselves and our property. The three kinds of NWS alerts are an ***advisory***, a ***watch***, and a ***warning***. It is important to understand what these words mean. A storm advisory means that a storm is anticipated. A storm watch means that a storm is expected. A storm warning means that a storm is in your area and you must take cover. You will learn the definitions for hazardous weather terms that may be involved in an NWS alert.

OBJECTIVE: Learn the definitions for common weather words.

Weather Tote

- Pencil

Do it

1. Find each word in the Word Bank in the word-search puzzle on the next page.
2. Match each word to the definition in the Weather Alerts Data Sheet as best as you can.
3. Discuss your answers with your teacher.

DEFINITIONS: *From the National Weather Service*

STORM ADVISORY An advisory is issued when a hazardous weather event is occurring, imminent, or likely. Advisories are for less serious conditions than warnings. Advisory conditions may cause significant inconvenience, and if caution is not exercised, could lead to situations that may threaten life or property.

STORM WATCH A watch is used when the risk of a hazardous weather event has increased significantly, but its occurrence, location, or timing is still uncertain. It is intended to provide enough lead time so those who need to set their plans in motion can do so. A watch means that hazardous weather is possible. People should have a plan of action in case a storm threatens, and they should listen for later information and possible warnings, especially when planning travel or outdoor activities.

STORM WARNING A warning is issued when a hazardous weather event is occurring, imminent, or likely. A warning means weather conditions pose a threat to life or property. People in the path of the storm need to take protective action.

WORD SEARCH PUZZLE

q	u	i	t	j	h	u	h	f	r	l	p
a	f	u	h	k	g	y	i	e	t	o	z
z	o	i	u	b	l	i	z	z	a	r	d
w	g	k	n	l	f	c	o	v	g	t	x
s	n	l	d	p	d	e	p	w	b	i	c
x	b	o	e	o	s	t	a	c	i	k	v
c	g	p	r	i	a	r	z	d	n	n	b
f	l	a	s	h	f	l	o	o	d	m	d
d	t	j	t	u	q	w	x	e	h	j	n
e	r	v	o	t	o	r	n	a	d	o	m
r	f	w	r	y	s	e	s	w	y	u	l
f	v	e	m	z	w	i	n	t	e	r	k

WORD BANK

blizzard
flash flood
fog
heat
ice
thunderstorm
tornado
wind
winter

WEATHER ALERTS DATA SHEET

Weather Alert	Kind of weather from Word Bank list
Winds of 58 mph or higher with hail of 1" in diameter or more with dangerous lightning	
Winds of 40 mph or higher or wind gusts of 58 mph or higher for one hour or more	
Imminent danger of rapid flooding in low-lying areas	
Blowing snow reducing visibility to a quarter mile or less for 3 hours or longer and sustained winds of 35 mph or greater or frequent gusts to 35 mph or greater	
A tornado is imminent	
¼ inch or more of ice accumulation	
The heat index value is expected to reach or exceed 110 degrees within the next 12–24 hours	
Reduced visibility to a quarter mile or less over a large area for an extended period (2 or more hours)	
A significant combination of hazardous winter weather is occurring or imminent	

SHARE WHAT HAPPENED:

- What are the NWS alerts?
- Which alert means that you need to take immediate action?

APPLY:

- Have you recently heard or seen a warning, watch, or advisory? If so, for what weather event was the warning, watch or advisory?
- Have you ever experienced a warning when you took action? If so, what was the action you took?

GENERALIZE TO YOUR LIFE:

- Does your family have a plan or preparations made for storm warnings? If so, what are the plans or preparations?
- Why might a family have a NOAA (National Oceanic and Atmospheric Administration) Public Alert Weather Radio?

Fly Higher

Create a plan for your family to prepare for a tornado warning and another plan for a winter storm warning.

NOTES

WEATHER OR CLIMATE?

What is the difference between weather and climate?

intro People often say climate and weather as if they mean the same thing. They do not. The temperature today is considered part of today's ***weather,*** but the average temperature over the past 30 years is considered part of our climate. Weather is the day-to-day changes in the climate. Weather is the day-to-day changes in the **atmosphere** and describes a single occurrence, such as the current temperature. An example of weather is a temperature reading of 85 degrees Fahrenheit on the 4th of July. An example of climate would be the average high temperature for your city or town on the 4th of July.

OBJECTIVE: Describe the difference between weather and climate.

Weather Tote

- Pencil
- Tables 4 and 5
- *Optional:* A resource on weather and climate, such as the Internet, a book, or a person

Do it

1. Imagine that you will spend the day outside today. Look or listen to today's weather report, and choose the clothes you will wear from the list of clothes given in Table 4, Dressing for Weather. Put an X next to the clothes that you choose to wear outside today.
2. You might hear any of the statements in Table 5 on your radio or TV. Read each statement in the first column of Table 5.
   - Indicate if the statement is an example of weather or climate by placing an X in one of the Example columns (2 or 3).
   - Explain your reason for your answer by placing an X in one of the Reason columns (4 or 5).

TABLE 4.
DRESSING FOR THE WEATHER

T-shirt		Shorts		Gym shoes	
Tank top		Jeans or long pants		Snow boots	
Long-sleeve shirt		Heavy socks		Light jacket	
Sweatshirt		Light socks		Raincoat	
Hoodie		Sandals		Winter coat	
Swimsuit		Flip-flops		Hat and gloves	

TABLE 5. WEATHER OR CLIMATE?

Announcer's Statement	Example		Reason	
	Weather	Climate	Single Occurrence	An Average
Yesterday the high was 55°F and the low was 43°F.				
Today we are expecting a high of 61°F.				
That is 10 degrees above the normal high of 51°F for this date.				
We will have clear skies today with no rain forecasted for the next three days.				
We usually would have four inches of rain this month.				
Hurricane season is beginning in the tropics.				
A tropical storm is developing in the Atlantic Ocean.				
To the north, a massive snowstorm is on its way.				
We do not normally see a snowstorm like this at this time of year.				
Tornado season is upon us, and we must be prepared.				
We usually expect four or five tornado outbreaks to occur this month.				
This morning a tornado damaged a building on High Street downtown.				

Chat: SHARE WHAT HAPPENED:

- Did you find it difficult to decide which of the statements in Table 5 were about weather and which were about climate?
- Are the clothes you wear today determined by the weather? Or the climate?
- Are the clothes you keep in your closet and dresser determined by the weather? Or the climate?
- Do you have clothes you don't wear very much? If so, why?

APPLY: Pick another time of year. What clothes from Table 4 would you wear outside at that time of year?

GENERALIZE TO YOUR LIFE:

- Would you keep all the same clothes if you lived in Hawaii? What would be different?
- Would you keep all the same clothes if you lived in Alaska? What would be different?

Fly Higher

Do you have a friend or relative who lives in another climate area—for example, Florida or Georgia; Arizona or California; Washington or Idaho; Maine or Vermont? Contact them, and compare the kind and number of types of clothes they keep and wear.

WHERE IS THE HEAT?

Why is it warmer in the summer than the winter?

Summers are warmer than winters. The amount of difference in the temperature partly depends on the climate **zone** that you live in—temperate, desert, tropical, or tundra. It also depends on your current weather conditions, including temperature, wind, cloudiness, precipitation, and **relative humidity**. We get heat from the sun, but how does the position of earth, in relation to the sun, affect the temperature? You will explore that relationship in this activity.

OBJECTIVE: Study how the angle of the sun affects earth's temperatures.

Weather Tote

- Flashlight
- Paper
- Ruler
- Pennies
- Pencil

PART 1

Question: If you shine a flashlight on the center of a piece of paper from two different places, will the shape of the lighted area stay the same?

1 Mark the center of a piece of paper by drawing an X from the corners of the paper.

2 Using the ruler as a guide, hold a flashlight 6 inches above the center of the paper, and turn on the flashlight.

3 Draw the shape of the light on the paper. Ask someone to hold the flashlight, if you need help.

4 Keep the flashlight 6 inches above your paper, and move the flashlight to the bottom of the paper. Shine the flashlight back toward the center of the paper.

5 Draw the shape of the newly lighted area.

6 Count the number of pennies you can fit in each of the shapes you drew.

PART 2

Question: How does the amount of daylight change over the year?

Calculate the total daylight hours for the first day of each season, using the data given in Table 6. Enter your calculation in the last column.

SHARE WHAT HAPPENED:

- What happened to the shape of the light when you moved the flashlight?
- How many pennies could you fit in each shape?
- How will the angle of the light affect the temperature?

WEATHER SKILLS:

- Which season has the shortest amount of daylight?
- Which season has the longest amount of daylight?
- Why would the amount of daylight make a difference in the seasons?
- Which season had the most direct sunlight?

APPLY: What effect do you think the angle of the sun has on temperature?

GENERALIZE TO YOUR LIFE: How does the amount and direction of sunlight your area receives affect what you wear in January? In June?

Fly Higher

- Compare the location of the sun when it rises and sets in January and July at your house.
- Compare the length of a shadow outside and close to your house at about noon during different seasons throughout the year.
- Do people living south of the equator have seasons at the same time we do? Do you know why? If not, research to find the answers.
- Ask someone to help you calculate the area of each shape you drew in Part 1. Are the areas different? If so, why do you think they are different?
- Graph the average length of daylight for Indianapolis, using the data in the Table 7.

TABLE 6.
DAYLIGHT HOURS

NOTE: Sunrise, sunset, and daylight hours data is from NOAA. Use the search box and keywords: sunrise, sunset, and daylight hours.

First Day of New Season	Sunrise	Sunset	Daylight (Hours)
Winter	8:06 a.m.	5:23 p.m.	
Spring	7:50 a.m.	8:00 p.m.	
Summer	6:17 a.m.	9:21 p.m.	
Fall	7:35 a.m.	7:46 p.m.	

TABLE 7. **INDIANAPOLIS'S DAYLIGHT HOURS**

Month	Daylight (hrs.)	Month	Daylight (hrs.)	Month	Daylight (hrs.)
January	9.7	May	14.4	September	12.4
February	10.7	June	14.9	October	11.1
March	11.6	July	14.6	November	10.0
April	13.3	August	13.7	December	9.4

GLOSSARY

Atmosphere: A layer of gases surrounding the earth and held in place by gravity.

Average annual extreme minimum temperatures: The sum of the lowest temperature each year for 30 years, divided by 30.

Condense: To change the state of a gas to a liquid

Dense: When the parts of a material are packed closer together, resulting in a heavy mass or weight.

Density: A measurement of the amount of space a material occupies.

Evaporate: To change the state of a liquid to a gas.

Hardiness: The ability of a plant to withstand cold temperatures.

Precipitation: The deposit on earth of forms of water, such as hail, mist, rain, sleet, and snow.

Relative humidity: The ratio of the amount of water vapor in the air compared to the amount of water vapor if that air is saturated, given as a percentage.

Water vapor: An invisible, colorless gas, like air; water in a gaseous state.

Zone: An area of similar temperature and precipitation conditions.

NOTES

NOTES

LEVEL 2

Weather and Climate Science Education for Grades 6–8

NOTE TO 4-H MEMBER

As you continue your study of weather and climate, you will be increasingly responsible for collecting your own data and gaining information from sources outside this manual. Discoveries from researching weather and earth's climates are often in the news. You will want to keep up to date with what scientists have recently reported.

Level 1 introduced you to basic weather terminology and concepts. Level 2 of *Weather and Climate Science Education for Grades 3–12* will introduce you to much more complex weather topics, such as air pressure, clouds, winds, humidity, and fronts. To better understand climate, you may choose to learn about earth's rotation and its connection to the high and low pressure systems that the weather reporter points out on the weather map. You may also choose to calculate your family's carbon footprint or study how what we do every day may create a "greenhouse effect" on earth's atmosphere that changes its temperature. The Level 3 will delve even deeper into weather and climate science concepts to help you fully understand weather and climate that you will use over your lifetime.

The key to learning in this, or any 4-H project, is to enjoy your studies and to learn at your own pace. We hope this project is just the start of a lifetime enjoyment of understanding the climate you live in and watching the weather!

Safety First!

Ask a parent or other adult to help when you see a safety first icon.

Knowing the Difference

Weather—The current weather conditions, including temperature, wind, cloudiness, precipitation, and relative humidity

Climate—The average weather over time

AIR PRESSURE

How does a change in air pressure affect liquid in a straw?

Air has weight because it is composed of tiny particles (molecules) in a gaseous state. The weight of those molecules creates air pressure. Air is pressing against you, even though you cannot see or feel it. This pressure does not press straight down, but all around you. The force pressing against you from every angle is, on average, 14.7 pounds, on every square inch of your body. This force is known as **air pressure**.

Most people understand that water is a **fluid** but don't realize that air is also a fluid. Air has the same fluid properties as water. Pressure in a fluid is caused by the weight of that fluid. The deeper you go into that fluid, the greater the pressure created by its weight. If you ever dived into a pool of water, you may have noticed an increase of pressure against your eardrums that is sometimes painful. This is caused by the increased weight of water above you. The force of this increased weight presses all around you, even on your ears.

Air pressure is greatest at the surface of the earth. Air pressure can indicate what kind of weather is coming. Changes in the **barometric pressure** indicate changes in the weather. When air pressure is rising, a *high pressure system* is on the way, possibly bringing cooler temperatures and clear skies. If a *low pressure system* is coming, look for relatively warmer temperatures and precipitation.

OBJECTIVE: Understand the nature of air pressure.

Weather Tote

- Clean glass jar (not plastic) with a metal screw-on lid (pint or similar size will work)
- Plastic straw, a clear skinny straw like some restaurants use
- Hot glue gun or Play-Doh
- Electric drill with a bit the same diameter as the straw

Follow these steps to construct a testing jar:

1. Drill a hole in the center of the metal lid that is the same diameter as the straw. Get help for this step, if you want, and make sure you drill the hole with the lid held on a block of thick wood so that you drill through the lid and into the wood.
2. Place the lid on the jar, and put the straw through the hole so that the bottom of the straw is about ½ inch from the bottom.

3. Seal the straw onto the lid, using a hot glue gun or Play-Doh. Seal it both inside and outside. Get help for this step, if you use the hot glue gun.

Test 1: Fill the jar to the brim with water and firmly place the lid on the bottle.

- Make sure no air bubbles are in the bottle. If bubbles appear, try again.
- Try to drink the water through the straw.
- Now try to blow air into the bottle.

Test 2: Take the lid off, and pour a little of the water out so about 1 inch of air is in the top of the bottle. Put the lid back on, and try again to drink the water. Drink as much water as you can without stopping. Notice what happens when you stop.

Now try to blow air into the bottle—do this part over a sink. Blow as much air as you can into the jar, and notice what happens after you stop.

SHARE WHAT HAPPENED:

Test 1:

- What happened as you tried to use the straw to drink the water?
- What happened when you tried to blow bubbles into the jar?
- Why do you think this happened?

Test 2:

- What happened as you tried to use the straw to drink the water?
- When you tried to drink the water, were you putting air into or taking air out of the straw?
- What happened to the water in the straw immediately after drinking some water?
- What happened when you tried to blow bubbles into the jar?
- When you tried to blow into the straw, were you putting air into or taking air out of the straw?
- What happened in the straw after you blew bubbles into the jar?

APPLY: When using a straw, differences in air pressure cause the water to move in the straw. In Test 1 of your experiment, no air pressure was in the jar, so there was no difference in air pressure to move the water in the straw in either direction. In Test 2 of your experiment, there was air in the jar, so you could create a difference in air pressure to push the water in the straw. Think about what happened in Test 2, and fill out the chart below.

Which direction does the water move? Circle one:

From low pressure to high pressure

From high pressure to low pressure

Action	Pressure in straw: high or low	Pressure in jar: high or low	Movement of water: up the straw or down the straw
Drinking some water from the straw			
Immediately after drinking some of the water			
Blowing air through the straw into the jar			
Immediately after blowing air into the jar			

GENERALIZE TO YOUR LIFE: Since air pressure is caused by the weight of air, where will air pressure be the greatest, at the base of a mountain or at the top of a mountain?

How can watching a change in air pressure help in predicting the weather?

In the atmosphere, does air move from high pressure toward low pressure or from low pressure toward high pressure?

Fly Higher

Use your testing jar as a ***barometer.*** Put about 2 inches of colored water in the jar. If the temperature around the jar stays the same, the jar will act as a barometer and the liquid in the straw will move as air pressure changes: pushing the liquid down with higher pressure and allowing the liquid to rise with lower pressure. You may need to blow a few bubbles of air into the jar to raise the level of the liquid within the straw. Keep a record of how this barometer changes with changes in air pressure by checking the pressure on a real barometer or a daily weather source.

[Did You Know? The value of 14.7 pounds of air pressure mentioned in the introduction is the **average** weight of 1 square inch of air measured from the earth's surface to the outer edge of our atmosphere.]

Notes

CARBON FOOTPRINTS

What is your family's carbon footprint? What can you do to reduce it?

Carbon dioxide (CO_2) is a greenhouse gas that is added to the atmosphere through many of our day-to-day activities such as driving cars and running gas-powered lawn mowers, and from power plants. The energy and materials that we use each day come from many sources. Some of these sources are **fossil fuels**. Fossil fuels contain carbon that is not part of the natural carbon cycle. The carbon has been stored in fossil fuels for a long time—in most cases, millions of years.

When we use fossil fuels, such as oil or natural gas, we are adding carbon dioxide to earth's atmosphere. Many products, including plastics, are made from or using fossil fuels, which also adds CO_2 to the atmosphere. Scientists have measured and recorded the atmosphere's CO_2 levels since 1958, and the levels continue to increase (see Figure 1). Carbon dioxide is a **greenhouse gas,** and the extra CO_2 in the air causes it to warm.

Whenever your family uses any of these energy sources or products, you have contributed to the additional carbon dioxide in the atmosphere. We call this your carbon footprint. The U.S. Environmental Protection Agency (EPA) has created a website to help you determine your family's carbon footprint.

OBJECTIVE: Determine your family's carbon footprint, and consider ways to reduce it.

FIGURE 1. **CARBON DIOXIDE LEVELS**

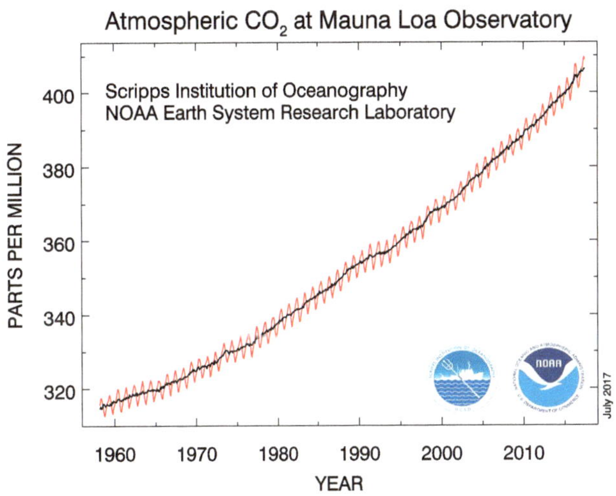

Source: National Oceanic and Atmospheric Administration | Department of Commerce

Weather Tote

- Access to the EPA website: www.epa.gov/climatechange/ghgemissions/ind-calculator.html
- Household information (good estimates, not exact totals):

  Average monthly energy bills

  For each vehicle your family owns, an estimated average miles driven per week and miles per gallon of each vehicle. To find each car's gas mileage, go to www.fueleconomy.gov/feg/findacar.shtml

FIGURE 2.
EPA CARBON FOOTPRINT WEBSITE

DO it

1 Go to the EPA Carbon Footprint website, shown in Figure 2. This page starts with a list of instructions. These instructions are under Instructions>Current Emissions>Reduce Emissions> Estimated Savings. Follow the instructions. To advance to the next section, click Next Section in the lower right corner. Follow through the calculator, using the instructions given in each section.

2 After entering your information, press your computer's tab key to get calculated results for each entry.

3 Enter the data in Table 8, Pounds of Carbon Dioxide Equivalent Emitted Per Year, by using the information on each page found in the right-hand column. Do not use the total in the Your Total Estimated Emissions (at the bottom), because those totals are a total of all the pages.

4 If you need to go back to a previous section, click on the icon at the bottom for Previous Section.

5 Complete Table 8, column 3.

6 Explore the Reduce Emissions section of the EPA webpage. See how much CO2 your family can save by choosing to do the following activities. Enter the value in column 2.

TABLE 8. **POUNDS OF CARBON DIOXIDE EQUIVALENT EMITTED PER YEAR**

	Family activity	**Total CO_2 (pounds)**
Household vehicles	Add the totals for each vehicle in the right-hand column.	
Home energy	Add the totals for each energy source in the right-hand column.	
Total waste emissions after recycling	Enter the total at the bottom of the right-hand column.	
Total	Add these three together, which totals the CO_2 from your family's actions.	
Estimated emissions	Enter your total from the bottom of the last screen. This total includes family size and location factors.	
The U.S. average for a household your size	Found in the bottom section, under your total estimated emissions.	

TABLE 9. **REDUCE EMISSIONS**

Action	CO_2 saved each year (pounds)
Reduce the miles driven by Vehicle 1 by 10 miles per week.	
Reduce the miles driven by Vehicle 2 by 10 miles per week.	
Turn down your household's heating thermostat by 2 degrees.	
Turn up your household's air-conditioner thermostat by 2 degrees.	
Replace five 60-watt incandescent bulbs with 13-watt energy-saving light bulbs—compact fluorescent bulbs.	
Describe one more action you and your family could take.	

Chat: SHARE WHAT HAPPENED:

- Do you contribute more carbon dioxide from your household's vehicles or from your home energy?
- How did your family's total compare with the U.S. average?
- How much total savings would you gain by the suggested actions?
- If 100 households in your neighborhood took these same actions, what would the total neighborhood reduction of carbon dioxide be per year?
- Would having a whole community follow these changes help reduce emissions?

APPLY: Make a list of actions you and your family can take to help reduce your carbon footprint.

How much carbon dioxide reduction will your actions create?

Ask your friends and relatives to take the carbon footprint test and compare your results. Are some people doing better than others? How?

GENERALIZE TO YOUR LIFE: How will your actions to reduce your family's carbon footprint affect your daily routine?

Fly Higher

Use online resources (.gov or .edu) or books on ecology or earth science to research other actions that you could take to help reduce your family's carbon footprint.

CLOUD FORMATION

How does air cool to form a cloud even on a warm, sunny day?

Water vapor in the atmosphere provides the moisture that makes clouds. Clouds form when air is cooled, which will cause water vapor to condense on particles in the air called aerosols. When the air becomes **saturated** (full of water), any additional water is shed and condenses into a cloud. This process is called **condensation**. Water vapor is an invisible gas, so if you see water in the air as a mist or cloud, you are really seeing water droplets or ice crystals, not water vapor.

More water vapor can exist in warm air than in cool air. When warm, moist air is cooled, the air becomes saturated, and some water vapor must condense out of the air to form droplets or ice crystals. There are many examples of air saturation at the surface of earth. Whenever warm, moist air is cooled by colder air, you can see a mist forming, like you do when you breathe out warm, moist air on a cold day, open a freezer door with warm, moist air outside, or see the air above a boiling pan of water cool the warm, moist air rising from the boiling water. Cold objects may cool the air to force saturation, like water condensing on a glass of ice water, dew forming on the cool grass, or frost forming on a cold windshield.

What causes cloud formation? To form a cloud, the air has to be cooled to cause condensation. Even on a warm, sunny day, clouds can form in the sky.

OBJECTIVE: Explain how cooling occurs in the atmosphere to form a cloud.

Weather Tote

- Two clear, empty 2-liter bottles—one bottle must be completely dry
- A Fizz-Keeper Pump Cap, which you can find at some grocery stores or online (Amazon.com)

Do it

1. Put about 2 inches of water in a 2-liter bottle, keeping the completely dry bottle for later.
2. Place the Fizz-Keeper Pump Cap securely on the bottle.
3. Shake the water in the bottle to cause as much evaporation as possible.
4. Pump air into the bottle.
   - Pump at least 100 times.
   - As you pump air into the bottle, you are increasing the air pressure and adding air that is relatively drier than the air in the bottle.
   - Shake the water some more after pumping to encourage more evaporation.
   - Let the bottle sit for at least 30 minutes, if possible, for maximum evaporation.

5 Unscrew the Fizz-Keeper Pump Cap from the bottle. Be careful to keep your hand on it **because it will pop off the bottle.**

6 Look for a cloud in the bottle.

- Chances are that you did not get a cloud flowing out of the bottle the first time, so try again.
- Add at least 25 more pumps each time you repeat the activity.
- Keep track of the number of pumps you use each time so that you can determine how many pumps will make the best cloud.
- The greater the pressure in the bottle, the better the cloud.

7 Place the Fizz-Keeper Pump Cap on the dry bottle, and try the experiment again with the number of pumps that made the best cloud in number 6.

QUESTIONS:

- How many pumps were necessary to get a cloud that flows out of the bottle?
- Why is it important to shake the water in the bottle and let the bottle sit?
- What is changing in the bottle when you pump air into it?
- What kind of results did you get with the dry bottle?
- Why is there a difference between your results with the dry bottle and the wet bottle?

DISCUSSION:

When you removed the Fizz-Keeper Pump Cap, the air in the bottle expanded. The air must be cooled to make a cloud. If air is forced to expand, that air will cool. Have you ever used a spray paint can? As the air expands out of the can, you feel the can getting colder. When you released the air pressure in the bottle, the expanding air cooled, causing the water vapor in that air to condense into a cloud. When you did the same thing using the dry bottle, you did not add any moisture to the air, so there was not enough water vapor for condensation to take place.

Air at the surface of earth is compacted by the higher air pressure at the surface. If that air is lifted up higher, where the air pressure is lower, it expands and cools, causing condensation to form a cloud. So lifted air causes most cloud formation. Fog and certain special clouds, such as mammatus, are exceptions and do not require lift to develop.

Look at the following diagrams that show ways air is lifted to form a cloud.

MAMMATUS CLOUDS Source: National Oceanic and Atmospheric Administration/Department of Commerce.

WAYS CLOUDS FORM:

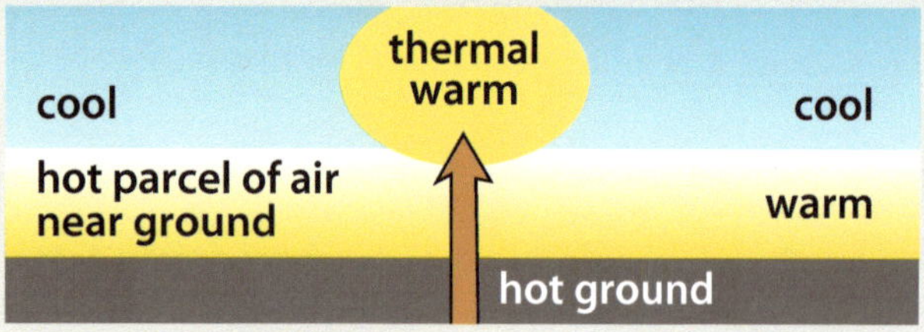

FIGURE 3. **THERMAL**
What is causing the air to lift? What is happening to the temperature of that rising air?

FIGURE 4. **MOUNTAIN EFFECT**
What is causing the air to lift? What is happening to the temperature of that rising air?

FIGURE 5. **COLD FRONT**
What is causing the air to lift? What is happening to the temperature of that rising air?

FIGURE 6. **WARM FRONT**
What is causing the air to lift? What is happening to the temperature of that rising air?

Have you ever seen a hawk or buzzard soaring in the sky? On sunny days these birds soar for long periods of time without flapping their wings! They often rise higher as they soar. Glider pilots look for small puffy clouds forming on a sunny day because they know by getting under them, they gain lift and stay aloft for a longer period of time.

Why can birds soar and even climb higher without flapping their wings?

SHARE WHAT HAPPENED: Which part of this activity did you enjoy the most? Why?

APPLY: How does the bottle and Fizz-Keeper Pump Cap activity relate to the four figures?

GENERALIZE TO YOUR LIFE: Why do glider pilots look for small puffy clouds on a sunny day?

Fly Higher

In this activity, you observed clouds formed by thermals. Any clouds present in the morning or after sunset are not there because of a thermal, but because of some frontal activity. On a clear, sunny day, watch for the formation of fair-weather cumulus clouds— mall puffy clouds. Check a weather report to see today's dew-point temperature. The higher the dew point, the more likely clouds will form. Watch at different times during the day to see how these clouds develop. Search for birds that might be riding the thermals. Observe these clouds in the evening—near, at, and after sunset. Record your observations in a journal, noting the time of day for each one.

CLOUD TYPES

How do we classify different kinds of clouds?

How would you describe the clouds you see in the sky today? Clouds come in all shapes and sizes. Some look puffy and white, while others have no form and seem dark and ominous. Clouds can tell us what is happening in the atmosphere and give us an idea of what kind of weather is approaching. Clouds may carry millions of tons of water or ice, and yet they appear to float in the sky. They bring fresh water to land through precipitation, so life can exist outside the ocean.

Five basic terms are used in naming clouds, which we combine to form the names of the ten basic clouds:

Cirrus or *cirro*—high-level, wispy or hair-like clouds

Stratus or *strato*—layered or flat clouds

Cumulus or *cumulo*—lumpy or fluffy clouds

Alto—midlevel clouds

Nimbus or *nimbo*—clouds holding any type of precipitation

The ten basic clouds are given below with their abbreviations. These clouds and a description of each can be seen on the NOAA website: https://www.noaa.gov/jetstream/clouds/ten-basic-clouds.

You will observe and identify the basic forms of clouds that might appear in your area over the course of several days.

OBJECTIVE: Observe and identify basic cloud types.

Weather Tote

- NOAA cloud images.

  Ten Basic Clouds on pages 52–54

  Cloud Classifications and Characteristics on pages 55–57
- Several days with clouds in the sky

HIGH-LEVEL CLOUDS	MID-LEVEL CLOUDS	LOW-LEVEL CLOUDS
Cirrus (Ci)	Altocumulus (Ac)	Cumulus (Cu)
Cirrocumulus (Cc)	Altostratus (As)	Cumulonimbus (Cb)
Cirrostratus (Cs)	Nimbostratus (Ns)	Stratocumulus (Sc)
		Stratus (St)

1 Make a cloud observation table with columns for the date, time, and your cloud observations. Make at least 10 rows under the column headers (date, time, observations).

2 Record the date, time, and name of the clouds you see in the sky by comparing the clouds that you see to those shown on the downloaded PDFs for 10 days.

3 Record all the cloud types you see each time you make an observation. You may see more than one type of cloud.

4 Record your data only on days when clouds are present.

5 You may record data up to three times each day, but the times you choose must be at least five hours apart.

NOTE: You may have trouble identifying the first cloud you observe. It is better to make a "best guess" than make no guess at all. You may see a cloud forming behind a high-flying airplane. Yes, that is actually a cloud, called a contrail, and you should include that in your observation sheet (this type has no abbreviation).

SHARE WHAT HAPPENED:

- How difficult was it to identify the clouds?
- Were there clouds that you were uncertain about? Which two types did you think each one could be?
- Did you see any clouds that did not fit any of the ten types?
- Did you observe any clouds with precipitation?

APPLY: When a cloud forms, the air is being lifted by some means. The lifting of the atmosphere can occur due to uneven heating of the surface or along a cold or warm front. A cold front is the boundary between where cold air is pushed into warm air. A warm front is the boundary between where warm air is being pushed over cooler air. Here are examples of those three types of lift.

WAYS CLOUDS FORM:

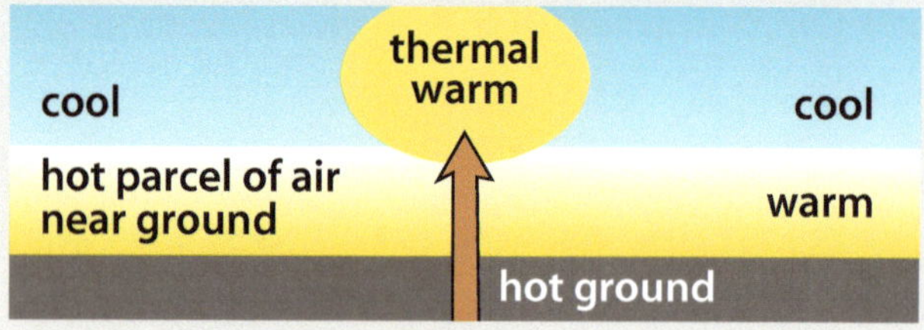

FIGURE 7. **UNEVEN HEATING**

FIGURE 8. **COLD FRONT**

FIGURE 9. **WARM FRONT**

- In which example will you find cumulonimbus clouds?
- In which example will you find fair-weather cumulus clouds?
- In which example will you find nimbostratus clouds?

GENERALIZE TO YOUR LIFE:

Clouds tell us about what is happening in the atmosphere. From your experience of watching and studying clouds, what kind of weather might you expect with the following types of clouds?

- Dark, thick stratus clouds
- Small, puffy cumulus clouds
- Towering dark cumulonimbus clouds
- High wispy cirrus clouds

Fly Higher

There are many other types of clouds than the ten basic cloud types you learned in this activity. Research other types of clouds, and expand your observations to see if you can discover any of them.

Take pictures of clouds, and create a photo gallery of your observations.

NOTES

TEN BASIC CLOUDS

From the World Meteorological Organization's (WMO) International Cloud Atlas, the official worldwide standard for clouds, the following are definitions of the ten basic cloud types.

John "Dr. Lightning" Jensenius, a NWS Meteorologist in Maine, created a PowerPoint Show "Cloud Book" called "The Clouds Outside My Window." It is available for free at https://www.noaa.gov/sites/default/files/2023-03/CloudsOutMyWindow.pdf. At the end of the 'book' is a link to a Power Point templet to make your own

High-Level Clouds

Cirrus (Ci)

Cirrocumulus (Cc)

Cirrostratus (Cs)

CIRRUS (Ci)

Detached clouds in the form of white, delicate filaments, mostly white patches or narrow bands. They may have a fibrous (hair-like) and/or silky sheen appearance. Cirrus clouds are always composed of ice crystals, and their transparent character depends upon the degree of separation of the crystals. As a rule when these clouds cross the sun's disk they hardly diminish its brightness. But when they are exceptionally thick they may veil its light and obliterate its contour. Before sunrise and after sunset, cirrus is often colored bright yellow or red. These clouds are lit up long before other clouds and fade out much later; some time after sunset they become gray. At all hours of the day cirrus near the horizon is often of a yellowish color; this is due to distance and to the great thickness of air traversed by the rays of light.

CIRROCUMULUS (Cc)

Thin, white patch, sheet, or layered of clouds without shading. They are composed of very small elements in the form of more or less regularly arranged grains or ripples. Most of these elements have an apparent width of less than one degree (approximately width of the little finger—at arm's length). In general cirrocumulus represents a degraded state of cirrus and cirrostratus both of which may change into it and is an uncommon cloud. There will be a connection with cirrus or cirrostratus and will show some characteristics of ice crystal clouds.

CIRROSTRATUS (Cs)

Transparent, whitish veil clouds with a fibrous (hairlike) or smooth appearance. A sheet of cirrostratus which is very extensive, nearly always ends by covering the whole sky. During the day, when the sun is sufficiently high above the horizon, the sheet is never thick enough to prevent shadows of objects on the ground. A milky veil of fog (or thin Stratus) is distinguished from a veil of Cirrostratus of a similar appearance by the halo phenomena which the sun or the moon nearly always produces in a layer of cirrostratus.

Mid-Level Clouds

Altocumulus (Ac)

Altostratus (As)

Nimbostratus (Ns)

ALTOCUMULUS (Ac)

White and/or gray patch, sheet or layered clouds, generally composed of laminae (plates), rounded masses or rolls. They may be partly fibrous or diffuse and may or may not be merged. Most of these regularly arranged small elements have an apparent width of one to five degrees (larger than the little finger and smaller than three fingers—at arm's length). When the edge or a thin semitransparent patch of altocumulus passes in front of the sun or moon a corona appears. This colored ring has red on the outside and blue inside and occurs within a few degrees of the sun or moon. The most common mid cloud, more than one layer of altocumulus often appears at different levels at the same time. Many times altocumulus will appear with other cloud types.

ALTOSTRATUS (As)

Gray or bluish cloud sheets or layers of striated or fibrous clouds that totally or partially covers the sky. They are thin enough to regularly reveal the sun as if seen through ground glass. Altostratus clouds do not produce a halo phenomenon nor are the shadows of objects on the ground visible. Sometime virga is seen hanging from altostratus, and at times may even reach the ground causing very light precipitation.

NIMBOSTRATUS (Ns)

Resulting from thickening altostratus, This is a dark gray cloud layer diffused by falling rain or snow. It is thick enough throughout to blot out the sun. The cloud base lowers as precipitation continues. Also, low, ragged clouds frequently occur beneath this cloud which sometimes merges with its base.

While altostratus and nimbostratus can extend into the high level of clouds, nimbostratus can extend into the low level as well.

Low-Level Clouds

Cumulus (Cu)

Cumulonimbus (Cb)

Stratocumulus (Sc)

Stratus (St)

CUMULUS (Cu)

Detached, generally dense clouds and with sharp outlines that develop vertically in the form of rising mounds, domes, or towers with bulging upper parts often resembling a cauliflower. The sunlit parts of these clouds are mostly brilliant white while their bases are relatively dark and horizontal. Over land cumulus develops on days of clear skies, and is due diurnal convection; it appears in the morning, grows, and then more or less dissolves again toward evening.

CUMULONIMBUS (Cb)

The thunderstorm cloud, this is a heavy and dense cloud in the form of a mountain or huge tower. The upper portion is usually smoothed, fibrous, or striated and nearly always flattened in the shape of an anvil or vast plume. Under the base of this cloud which is often very dark, there are often low ragged clouds that may or may not merge with the base. They produce precipitation, which sometimes is in the form of virga. Cumulonimbus clouds also produce hail and tornadoes.

Note: Cumulus and cumulonimbus clouds are usually found in the low level category, but their tops may reach into the mid and high levels.

STRATOCUMULUS (Sc)

Gray or whitish patch, sheet, or layered clouds, which almost always have dark tessellations (honeycomb appearance), rounded masses, or rolls. Except for virga they are nonfibrous and may or may not be merged. They also have regularly arranged small elements with an apparent width of more than five degrees (three fingers—at arm's length).

STRATUS (St)

A generally gray cloud layer with a uniform base which may, if thick enough, produce drizzle, ice prisms, or snow grains. When the sun is visible through this cloud, its outline is clearly discernible. Often when a layer of stratus breaks up and dissipates blue sky is seen. Sometimes appearing as ragged sheets stratus clouds do not produce a halo phenomenon except, occasionally at very low temperatures.

CLOUD CLASSIFICATIONS AND CHARACTERISTICS

By Ted Funk, Meteorologist in Charge, National Weather Service

Clouds are classified according to their height above and appearance (texture) from the ground. The following cloud roots and translations summarize the components of this classification system: 1) Cirro-: curl of hair, high; 2) Alto-: mid; 3) Strato-: layer; 4) Nimbo-: rain, precipitation; and 5) Cumulo-: heap.

High-level Clouds

High-level clouds occur above about 20,000 feet and are given the prefix "cirro." Due to cold tropospheric temperatures at these levels, the clouds primarily are composed of ice crystals, and often appear thin, streaky, and white (although a low sun angle, e.g., near sunset, can create an array of color on the clouds). The three main types of high clouds are cirrus, cirrostratus, and cirrocumulus.

Cirrus clouds are wispy, feathery, and composed entirely of ice crystals. They often are the first sign of an approaching warm front or upper-level jet streak. Unlike cirrus, cirrostratus clouds form more of a widespread, veillike layer (similar to what stratus clouds do in low levels). When sunlight or moonlight passes through the hexagonal-shaped ice crystals of cirrostratus clouds, the light is dispersed or refracted (similar to light passing through a prism) in such a way that a familiar ring or halo may form. As a warm front approaches, cirrus clouds tend to thicken into cirrostratus, which may, in turn, thicken and lower into altostratus, stratus, and even nimbostratus.

Finally, cirrocumulus clouds are layered clouds permeated with small cumuliform lumpiness. They also may line up in "streets" or rows of clouds across the sky denoting localized areas of ascent (cloud axes) and descent (cloud-free channels).

Mid-level Clouds

The bases of clouds in the middle level of the troposphere, given the prefix "alto," appear between 6,500 and 20,000 feet. Depending on the altitude, time of year, and vertical temperature structure of the troposphere, these clouds may be composed of liquid water droplets, ice crystals, or a combination of the two, including supercooled droplets (i.e., liquid droplets whose temperatures are below freezing). The two main type of mid-level clouds are altostratus and altocumulus.

Altostratus clouds are "strato" type clouds (see below) that possess a flat and uniform type texture in the mid levels. They frequently indicate the approach of a warm front and may thicken and lower into stratus, then nimbostratus resulting in rain or snow. However, altostratus clouds themselves do not produce significant precipitation at the surface, although sprinkles or occasionally light showers may occur from a thick alto-stratus deck.

Altocumulus clouds exhibit "cumulo" type characteristics (see below) in mid levels, i.e., heaplike clouds with convective elements. Like cirrocumulus, altocumulus may align in rows or streets of clouds, with cloud axes indicating localized areas of ascending, moist air, and clear zones between rows suggesting locally descending, drier air. Altocumulus clouds with some vertical extent may denote the presence of elevated instability, especially in the morning, which could become boundary-layer based and be released into deep convection during the afternoon or evening.

Low-level Clouds

Low-level clouds are not given a prefix, although their names are derived from "strato" or "cumulo," depending on their characteristics. Low clouds occur below 6,500 feet, and normally consist of liquid water droplets or even supercooled droplets, except during cold winter storms when ice crystals (and snow) comprise much of the clouds.

The two main types of low clouds include stratus, which develop horizontally, and cumulus, which develop vertically. Stratus clouds are uniform and flat, producing a gray layer of cloud cover which may be precipitation-free or may cause periods of light precipitation or drizzle. Low stratus decks are common in winter in the Ohio Valley, especially behind a storm system when cold, dismal, gray weather can linger for several hours or even a day or two. Stratocumulus clouds are hybrids of layered stratus and cellular cumulus, i.e., individual cloud elements, characteristic of cumulo type clouds, clumped together in a continuous distribution, characteristic of strato type clouds. Stratocumulus also can be thought of as a layer of cloud clumps with thick and thin areas. These clouds appear frequently in the atmosphere, either ahead of or behind a frontal system. Thick, dense stratus or stratocumulus clouds producing steady rain or snow often are referred to as nimbostratus clouds.

In contrast to layered, horizontal stratus, cumulus clouds are more cellular (individual) in nature, have flat bottoms and rounded tops, and grow vertically. In fact, their name depends on the degree of vertical development. For instance, scattered cumulus clouds showing little vertical growth on an otherwise sunny day used to be termed "cumulus humilis" or "fair weather cumulus," although normally they simply are referred to just as cumulus or flat cumulus. A cumulus cloud that exhibits significant vertical development (but is not yet a thunderstorm) is called cumulus congestus or towering cumulus. If enough atmospheric instability, moisture, and lift are present, then strong updrafts can develop in the cumulus cloud leading to a mature, deep cumulonimbus cloud, i.e., a thunderstorm producing heavy rain. In addition, cloud electrification occurs within cumulonimbus clouds due to many collisions between charged water droplet, graupel (ice-water mix), and ice crystal particles, resulting in lightning and thunder.

Cirrus clouds (above)

Cirrostratus clouds (above)

Cirrocumulus clouds (above)

Altostratus clouds (above)

Altocumulus clouds (above)

Stratus clouds (above)

Stratocumulus clouds (above)

Other Interesting Clouds

Wall Cloud: A localized lowering from the rain-free base of a strong thunderstorm. The lowering denotes a storm's updraft where rapidly rising air causes lower pressure just below the main updraft, which enhances condensation and cloud formation just under the primary cloud base. Wall clouds take on many shapes and sizes. Some exhibit strong upward motion and cyclonic rotation, leading to tornado formation, while others do not rotate and essentially are harmless. Shelf Cloud: A low, horizontal, sometimes wedge-shaped cloud associated with the leading edge of a thunderstorm's outflow or gust front and potentially strong winds. Although often appearing ominous, shelf clouds normally do not produce tornadoes. Fractus: Low, ragged stratiform or cumuliform cloud elements that normally are unattached to larger thunderstorm or cold frontal cloud bases. Also known as scud, fractus clouds can look ominous, but by themselves are not dangerous. Mammatus: Drooping underside (pouchlike appearance) of a cumulonimbus cloud in its latter stage of development. Mammatus most often are seen hanging from the anvil of a severe thunderstorm, but do not produce severe weather. They can accompany nonsevere storms as well. Contrail: Narrow, elongated cloud formed as jet aircraft exhaust condenses in cold air at high altitudes, indicative of upper level humidity and wind drift. Fog: Layer of stratus clouds on or near the ground. Different types include radiation fog (forms overnight and burns off in the morning) and advection fog.

Nimbostratus clouds (above)

Fractus cloud (above)

Cumulus clouds (above)

Fog (above)

Cumulus congestus (above)

Contrails (above)

Cumulonimbus (above)

Mammatus cloud (above)

Wall cloud (above)

Shelf cloud (above)

COLD FRONTS

What is a cold front, and what kind of weather does a cold front produce?

intro Across many parts of the country, a cold front often comes before precipitation (rain or snow), although a cold front can move though without precipitation. Cold fronts are marked by a shift in wind. A front is a boundary between two air masses and occurs when one air mass moves into the area of another air mass. The difference between the air masses is their temperatures, which creates a change in air density. A cold front is a large boundary of air—generally extending from north to south when fully developed in the Midwest—with colder air behind the front and warmer air ahead of the front. This front, usually moving eastward in the Midwest, can trigger clouds to form, resulting in precipitation. In this activity, you will simulate your own cold front.

OBJECTIVE: Understand the nature of a cold front.

Weather Tote

- Two identical baby food jars or any two wide-mouth jars with identical rims. You do not need the lids.
- Waxed paper or an index card
- Blue food coloring
- Red food coloring
- Water
- Baking pan

Do it Complete this activity in the baking pan to catch spilled water. Place the baking pan near a window or light so that you can easily see what happens in the jars.

1 Place one jar upside down on the other jar to ensure that they are an exact match. If the jars do not match, find a different pair.

2 Fill Jar 1 with cold water.

3 Add one or two drops of blue food coloring to the cold water. Stir the mixture.

4 Add more cold water so that Jar 1 is full to the rim.

5 Fill Jar 2 with hot tap water to about 1 inch below the rim.

6 Add one or two drops of red food coloring to the hot water, and stir the mixture.

7 Add more hot water so that Jar 2 is full to the rim.

8 Place a piece of waxed paper or an index card on the top of the hot-water jar, and press down around the edges of the jar to make a seal.

9 While keeping your hand flat on the paper, slowly turn over the jar until it is upside down in your other hand. Air pressure will hold the paper in place.

10 Place the sealed top of Jar 2 (hot water) on the top of Jar 1 (cold water), so that the rims match. The paper on Jar 2 will act as a boundary between the two jars.

11 Once the jars are stacked on each other, slowly and carefully remove the paper between the jars by pulling gently, while keeping your hand on the two jars to keep the rims matched up. You may need a helper with this part. Do not worry if the paper begins to fall apart and a little water spills out.

12 Carefully lift the two joined jars, keeping the rims together, and slowly tip the jars 90° to the right. Jar 1 (cold water) should be on the left, and Jar 2 (warm water) should be on the right. Keep the jars above the pan, and keep one hand on each jar.

13 Observe how the colored water moves in the jars.

SHARE WHAT HAPPENED:

- What did you observe as the jars were tipped to the right?
- Why did the cold water affect the warm water in this way?

APPLY: Use Figure 10 to answer the following questions:

- How similar is the diagram of a cold front, in Figure 10, to what you observed with the two colored waters in your jars?
- Why are clouds forming at the boundary between the two air masses?

GENERALIZE TO YOUR LIFE: You see an approaching cold front on a weather map. What kind of weather changes are you most likely to observe when the front arrives? Why?

This activity adapted from: https://www.thoughtco.com/simulate-weather-fronts-3444312

FIGURE 10. **COLD FRONT**

EARTH'S ROTATION

How does the rotation of the earth affect the movement of air across its surface?

The ideas of French physicist Gaspard Coriolis were applied in the early 1900s to the movement of air to explain the rotation of wind in high and low pressure systems. In 1835 he had described the force that results when an object moves across a rotating surface. As earth rotates, objects on its surface at different latitudes move at different speeds. Depending on the direction of earth's rotation, the air turns from its straight path. This force that deflects air, or wind, away from a straight course is called the Coriolis effect.

OBJECTIVE: Identify the cause and effect of earth's rotation on wind movement.

Weather Tote

- A rotating globe (or a ball with a mark for the North and South poles if you do not have a globe)
- Masking tape
- Pen
- A clock with second, minute, and hour hands
- Tennis ball (or any small soft ball)
- The Coriolis worksheet
- Access to a playground roundabout (sometimes called a merry-go-round) at a schoolyard or park
- Two friends to help you with this experiment at the playground

Rotation 1

1 Place two pieces of masking tape (about 1") on the globe—one just above the equator (Northern Hemisphere) and one just below the equator (Southern Hemisphere).

2 Turn the globe slowly counterclockwise while looking down at the North Pole.

- Draw an arrow on the tape in the Northern Hemisphere showing the direction it moves.

3 Turn the globe over so you are now looking down at the South Pole.

4 Rotate the globe in the direction of your arrow in the Northern Hemisphere.

- Draw an arrow on the tape in the Southern Hemisphere showing the direction it moves.

5 Answer the Rotation 1 questions on the worksheet on pages 62 and 63.

Timing

8 Ask a friend to help you while you study the clock.

9 Watch the minute hand for one minute.

10 Answer the question in the data sheet.

11 Watch the hour hand for one minute.

12 Answer the Timing questions on the worksheet.

Position

13 Calculate the speed of earth at different latitudes in the Coriolis data sheet.

14 Answer the Position questions on the worksheet.

Rotation 2

15 Visit a roundabout with two friends, and take a small ball.

16 Ask your friends to do the following:

Sit (do not stand) on the roundabout opposite each other.

Practice tossing the ball back and forth.

17 Turn the roundabout slowly counterclockwise, while your friends continue to toss the ball back and forth five times.

18 Answer the questions on the worksheet on page 63.

19 Turn the roundabout slowly clockwise, while your friends continue to toss the ball back and forth five times.

20 Answer the Rotation 2 questions on the worksheet.

SHARE WHAT HAPPENED: Were you surprised about any of your findings?

APPLY: Have you ever walked between surfaces that are moving at different speeds? For example, stepping onto a moving sidewalk at an airport or amusement park ride? How does that affect your balance?

How is wind affected in the Northern Hemisphere, which is rotating counterclockwise? Is the wind being deflected to the right or left?

GENERALIZE TO YOUR LIFE: What objects might be affected by the Coriolis effect?

Fly Higher

Explore other objects that move across earth's surface and might be affected by the Coriolis effect, such as airplanes and missiles. The Internet may be your best resource.

EARTH'S ROTATION WORKSHEET

Rotation 1

What do you notice about the two arrows? ______________________________

____

____

What is the difference in rotation between the Northern and Southern hemispheres?

____

____

Timing

Did you see the minute hand move while you were watching it? ________________

Did you see the hour hand move while you were watching it? __________________

The hour hand on a clock rotates once every 12 hours, and it takes earth 24 hours to rotate once. So the hour hand rotates ______________________________ as fast as earth.

Position

Calculate the speed of earth at different latitudes: Speed = distance ÷ by time (24 hours). Round to the nearest mile (no decimals).

Location (Latitude)	Distance Traveled (in 24 hours)	Speed (miles/hour)
Equator (0°)	About 25,000 miles	
Indianapolis, Indiana (about 40°)	About 18,000 miles	
Anchorage, Alaska (about 60°)	About 12,000 miles	

Answer the following questions by comparing the speeds you calculated above.

What part of the earth travels the fastest? (circle one)

Equator   Indianapolis   Anchorage

Is Chicago rotating   faster   or   slower   (circle one) than you?

Is New Orleans rotating   faster   or   slower   (circle one) than you?

Rotation 2

What happened when you started rotating the roundabout counterclockwise (similar to the motion of the Northern Hemisphere)?

____

____

Did the ball seem to curve? (circle one)   Yes   No

Which way? (circle one)   No curve   Right   Left

What happened when you started rotating the roundabout clockwise (similar to the motion of the Southern Hemisphere)?

Did the ball seem to curve? (circle one)   Yes   No

Which way? (circle one)   No curve   Right   Left

It is likely that one of your friends will need to retrieve the ball until they get used to tossing and catching it. Watch the apparent motion of the ball. Note how the ball appears to curve (right or left).

When your friend turns the roundabout slowly clockwise it is like the rotation of the Southern Hemisphere. Note how the ball appears to curve (right or left).

Coriolis described the force that results when an object moves across a rotating surface. Because earth rotates, objects on earth's surface at different latitudes move at different speeds and can be deflected. This force, which moves air, or wind, away from a straight course, is called the Coriolis effect.

How do you think the earth's rotational speed affects wind and weather patterns?

____

____

GLOBAL WINDS

How do the sun's energy and earth's rotation combine to create global wind patterns?

intro While we may experience winds blowing from any direction on any given day, the weather systems in the Midwest usually travel from west to east. People in Indiana can look at Illinois weather to get an idea of what to expect the next day. This predictable weather pattern occurs because the Midwest is in the middle latitudes where the prevailing westerlies, the region's normal winds, push the weather from west to east. The earth has definite wind patterns that are a result of how the sun heats a rotating earth. These patterns influence the different types of climate around the world.

OBJECTIVE: Describe the global wind patterns and how these patterns are a result of the sun's energy and the earth's rotation.

Weather Tote

- Global Winds worksheet

Do it

1. Read the Global Winds Information.
2. Complete the Global Winds worksheet on pages 67–69.

GLOBAL WINDS INFORMATION

Air that moves across the surface of earth is called wind. The sun heats the earth's surface, which warms the air above it. Areas near the equator receive the most direct sunlight and warming. The North and South poles receive the least amount of sunlight so these areas are cooler. More water vapor can be present in warmer air. You have probably experienced very warm weather and high humidity (moisture in the air) in the summer. Cooler air is denser (heavier) than warmer air, which is often described as "warm air rises." This phrase often refers to heat rising from the floor toward the ceiling in a room, or from lower floors to upper floors in a house or apartment. This movement is also true globally. The surface at the equator is both warm and humid. The cold, dry air at the North and South poles is much denser than the warm, humid air near the equator. So warm air tends to move northward in the Northern Hemisphere and southward in the Southern Hemisphere.

WIND PATTERNS

If global winds were affected only by the heating of the sun, they would move only toward the poles. The surfaces near the equator would become very hot and humid, causing that air to rise and flow toward the North and South poles. The cooled air would sink and return to the region of the equator. But earth is always rotating, which prevents air from traveling in a straight line. The earth's rotation forces the air to the right in the Northern Hemisphere and to the left

GLOBAL WIND PATTERNS

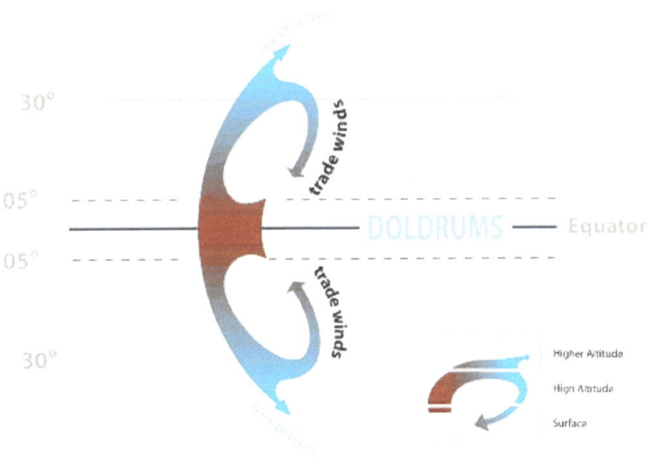

Source: Source: National Oceanic and Atmospheric Administration/Department of Commerce.

in the Southern Hemisphere. But rather than one large global wind from the equator to each pole, there are six bands of circulation which fluctuate with the seasons—three in the Northern Hemisphere and three in the Southern Hemisphere. These regions, called cells, are described below for the Northern Hemisphere. The same patterns exist in the Southern Hemisphere.

- The Hadley cell region is named after George Hadley (1685–1768), an amateur meteorologist who suggested that surface winds existed. The air rises at the equator (0°) and sinks at about 30° north latitude, moving back toward the equator at the surface. These surface winds are called trade winds, and they generally blow from east to west. Hurricanes develop within the trade winds.
- In the Ferrel cell region, suggested by William Ferrel (1817–1891), the air near the surface at 30° north latitude sinks as it also moves northward. Earth's rotation deflects this northward surface air to the right, and these surface winds begin to blow generally from west to east. In the area between 30° north latitude and 60° north latitude, the surface winds are known as the prevailing westerlies—sometimes called the stormy westerlies.
- The third latitudinal cell is known as the polar cell region. Here, air rises at 60° north latitude and moves toward the North Pole (90°), where the air cools and sinks. This results in surface winds that move generally from east to west. These winds are called the polar easterlies. The sinking air creates polar deserts, but they are not like the deserts that you may have seen or heard about in hot regions of the world. The North and South poles do not get large amounts of precipitation, but the snow they do receive builds up year after year into deep snow packs.

This creates what seems like a uniform global system, but several factors combine to make it more complicated. Factors that affect the weather include landmasses, ocean currents, and seasonal changes, which cause otherwise orderly global wind patterns to fluctuate and make predicting weather so complicated.

HIGH- AND LOW-PRESSURE REGIONS

Areas of high and low pressure exist at earth's surface between each of the cell regions. They are described below.

0° latitude—the doldrums: A band of low pressure runs along the equator between the north and south trade winds. Very little surface wind is present within this band of lifting air. For this reason, sailing ships often got caught in this area and spent days waiting for enough wind to carry them out. This area became known as the doldrums, as ships slowed and floated aimlessly.

30° latitude—the horse latitudes: Along 30° N and S latitude between the trade winds and the prevailing westerlies is an area of high pressure. Within this band of sinking air is very little surface wind. As in the doldrums, sailing ships got caught in this area and spent days waiting for enough wind to carry them out. In the centuries following Christopher Columbus's voyage, sailing ships crossed this band of high pressure on their way to what is now called the Americas. They carried people and supplies, along with horses, which were essential to the settlers but not native to the Americas. When sailing ships got caught in these windless areas and supplies began to disappear, the horses became dispensable and the travelers abandoned them. For this reason, this band became known as the horse latitudes.

Because air is sinking along this band and clouds do not form in sinking air, the major deserts of the world can be found at 30° latitude. Deserts form in areas where air is normally sinking, either along 30° latitude, north or south, or on the leeward side of mountains, which typically receives little precipitation or wind.

At 60° latitude—the subpolar lows: Along the 60° latitudes between the prevailing westerlies and the polar easterlies is a band of low pressure. This area also has no surface winds.

Chat

SHARE WHAT HAPPENED: What force causes air to move?

APPLY: Name a city, town, or county that usually gets the same weather you observe in your home's area after you experience it.

GENERALIZE TO YOUR LIFE: How can understanding global wind patterns help you prepare for the next few days' weather?

Fly Higher

Look at a world map or globe. Find all of the major deserts, and explain how sinking air creates each desert.

NOTES

GLOBAL WINDS WORKSHEET

How might the sun have an effect on global wind patterns?

____

____

How might the rotation of the earth affect global wind patterns?

____

____

Why is the air less dense at the surfaces near the equator than those at the North and South poles?

____

____

Why is warm, moist air rising above the surfaces near the equator?

____

____

In what direction is the surface air shown on page 65 moving—toward areas near the equator or those away from the equator?

____

____

In what direction is the air high above the surface moving—toward areas near the equator or those away from the equator?

____

____

Cell Region	Location in Latitudes	Name of Surface Winds	Direction of Winds
Hadley			
Ferrel			
Polar			

Fill in the table above.

What prevents simple global surface winds from blowing from the North and South poles to the equator?

____

____

Describe how air moves in the Hadley cell region:

____

____

Describe how air moves in the Ferrel cell region:

____

____

Describe how air moves in the polar cell region:

____

____

Why is the area along the equator known as the doldrums?

____

____

Why do deserts form along 30° latitudes?

____

____

Indicate the locations of the following in the figure below.

- Doldrums
- Horse latitudes
- Subpolar lows

Show the general wind direction for the following regions in the figure below:

- Polar easterlies
- Prevailing westerlies
- Trade winds

LATITUDE LINES (30 AND 60)

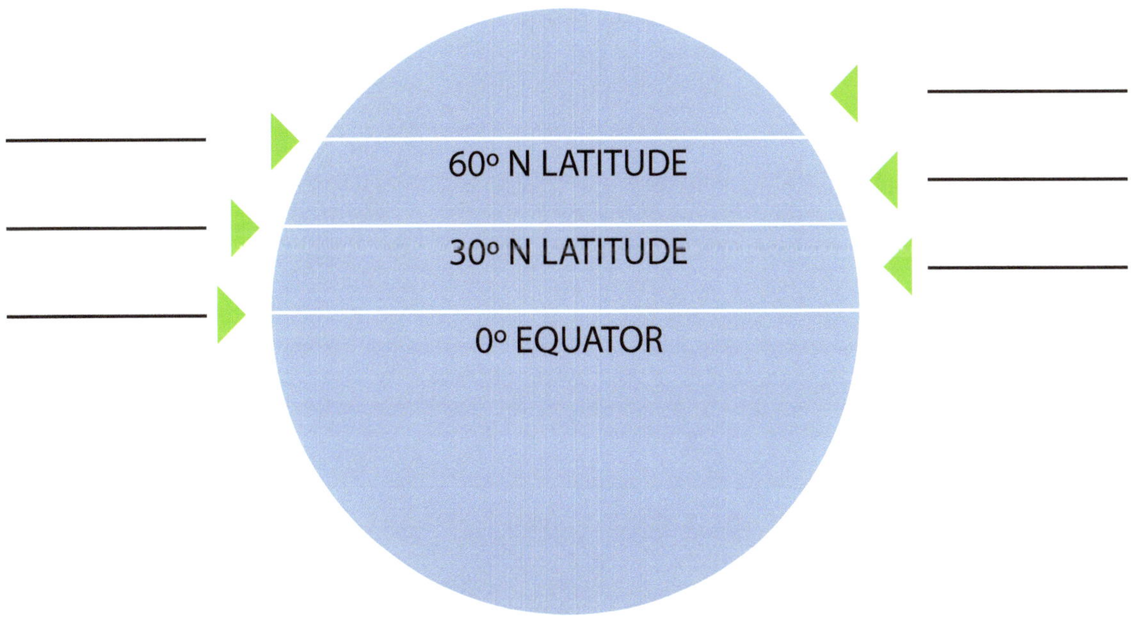

THE GREENHOUSE EFFECT

What is the greenhouse effect, and what impact does it have on earth's atmosphere?

A greenhouse is used to grow plants—sunlight passes through the glass of a greenhouse, and plants inside absorb the sun's rays. The sun's rays warm plants and other objects in a greenhouse. The warmed plants, floors, benches, and other objects give off heat energy. But this heat energy does not pass back through the greenhouse glass very well. Some stays in the greenhouse, making the interior warm. Because of this, greenhouses can be used to grow plants in cold weather.

Earth's atmosphere acts a little like a huge greenhouse, so this analogy is often used to help people understand how heat is trapped. Figure 11 shows the atmospheric "greenhouse effect."

The sun supplies the heat energy that warms earth, but sunlight does not warm the atmosphere, because it doesn't absorb light energy. Objects on earth's surface, on the other hand, absorb the light energy and become warm. They, in turn, give off heat energy, called infrared radiation or longwave radiation.

Most of this heat energy will pass right through the atmosphere and back into space. Some gases in the atmosphere, such as carbon dioxide, water vapor, nitrous oxide, and methane, absorb heat energy. These gases are called greenhouse gases because they act like the glass in a greenhouse and keep some of the heat energy from escaping. Without these greenhouse gases, earth would be a much colder planet.

FIGURE 11. **GREENHOUSE EFFECT**

Do you think that changes in the concentration of greenhouse gases could have an effect on the temperature of the atmosphere?

OBJECTIVE: Describe the greenhouse effect, and demonstrate its ability to increase the temperature of earth's atmosphere.

Weather Tote:

- A sunny day with calm winds or a desk lamp with a 60- or 75-watt bulb
- Two shoeboxes, two large glass jars, or similar containers. Containers should be the same size. Lids are not required.

- Two thermometers that will fit in the containers
- Cardboard to make two thermometer stands
- Clear plastic wrap or a clear plastic bag large enough to hold the containers.
- A watch or a clock with a second hand
- A ruler and a pencil
- Potting soil, enough for about a 1-inch layer in each container
- Two small labels or sticky notes
- Graph or grid paper

Predict the following:

Current temperature in the sun or room where you will complete the experiment:

____

Final temperature in an open container:

____

Final temperature in a container sealed in plastic:

____

1 Put an inch of dark soil in the bottom of the containers.

2 Attach the thermometers to pieces of cardboard so the thermometer bulb is not resting on the soil, but is about an inch above.

3 Place one cardboard-mounted thermometer on the soil inside each container. Position the thermometers so that they can easily be read during the experiment.

4 Cover one container with plastic wrap to contain greenhouse gases. Make sure the container is sealed well.

5 Label the containers:

Covered container: A

Open container: B

6 Record the temperature on each thermometer. They should be the same temperature. If they are not, you will need to adjust one. For example, if the thermometer in container B is one degree warmer than the initial reading for the thermometer in container A, you will need to subtract one degree from all your readings of thermometer B (calibration).

7 Initial temperature: ______________________

8 Place the containers next to each other in direct sunlight or under a lamp. If you are doing this experiment inside, position the lamp about 6 inches above the containers so the light spreads evenly between them.

9 Record the temperature of each container every five minutes for 30 minutes while the containers are in sunlight or under the lamp.

10 Remove the light source: take the containers inside or turn the light off.

11 Record the temperatures every five minutes for 30 minutes.

- Create a line graph with your data.
- Enter the time on the x-axis.
- Enter the temperature range and units on the y-axis.
- Enter the temperature data points you collected.
- Connect the data points.

TEMPERATURE DATA

With Light Source			Without Light Source		
Time (minutes)	Temperature (°F or °C) A	B	Time (minutes)	Temperature (°F or °C) A	B
0			0		
5			5		
10			10		
15			15		
20			20		
25			25		
30			30		

SKETCH YOUR GRAPH BELOW:

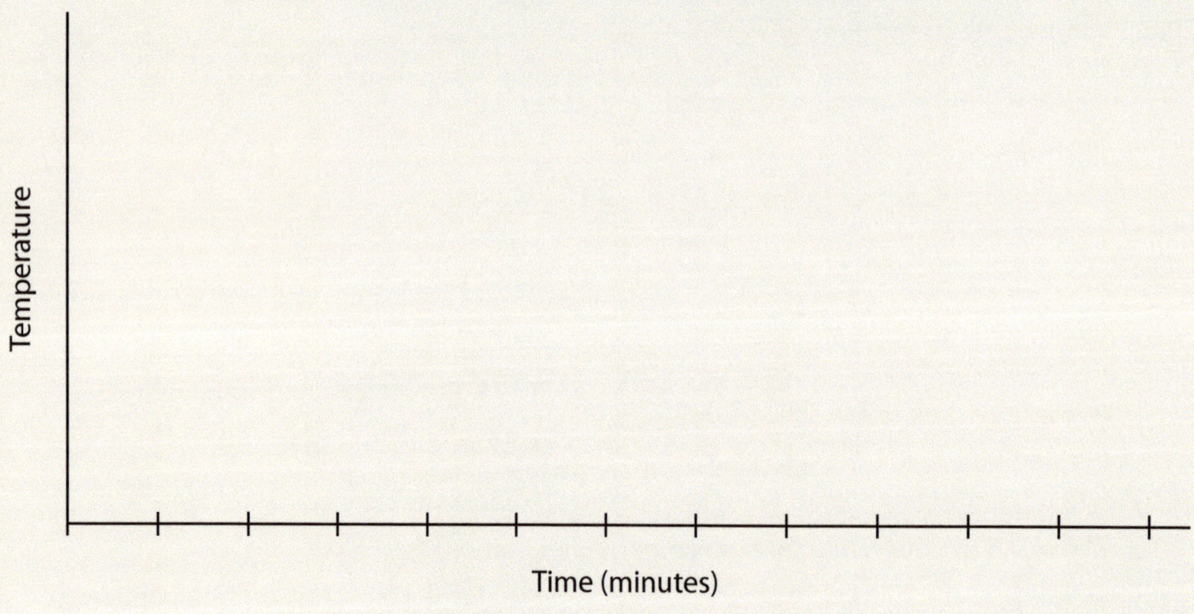

SHARE WHAT HAPPENED:

- Were your predictions correct?
- Which container had the higher temperature after the first 30 minutes—A or B?
- What caused this difference?
- After the second 30-minute period, which container had the higher temperature?
- What caused this difference?
- How is the plastic wrap like greenhouse gases?

APPLY: In March 1958, C. David Keeling, a 30-year-old scientist, began to keep a record of the amount of carbon dioxide present in the atmosphere with a method he devised as a PhD student. Other scientists during the 1950s suspected that human activity might be influencing the amount of this important greenhouse gas, but no one had a way to measure it until Keeling. He decided to do his atmospheric measurements in an area far from the influences of major industrial areas, so he went to Mauna Loa in Hawaii.

Figure 12 displays the carbon-dioxide data collected from March 1958 through July 2014. Study the graph. Then, describe how you think this steady increase in carbon dioxide, an important greenhouse gas, is affecting the temperature of earth's atmosphere.

FIGURE 12. **ATMOSPHERIC CO_2 AT MAUNA LOA OBSERVATORY**

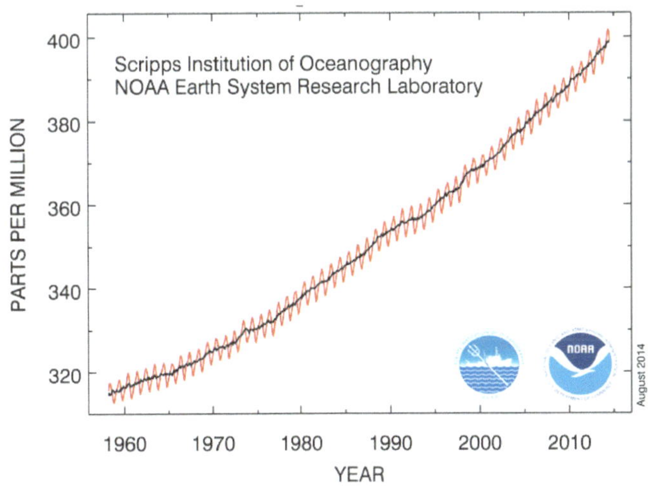

Source: National Oceanic and Atmospheric Administration | Department of Commerce

GENERALIZE TO YOUR LIFE: What is the effect of the increase in atmospheric CO2 on the earth?

Fly Higher

A car acts just like a greenhouse. The windows let sunlight in but trap the heat energy. Experiment with your parent's car on a sunny, hot day by recording the temperature inside the car under different conditions—in direct sunlight, in the shade, with the windows open at different levels, using a windshield shade, and any other ideas you might have. You will need two thermometers: one inside the back seat and one outside near the car but not placed on a hot object. Always keep your thermometers out of direct sunlight. Record your data in both table and graph form.

HUMIDITY

How does moisture in the air affect how you feel?

intro Humidity refers to the amount of water vapor in the air. Hot air can hold much more water vapor than cold air. When the air has a lot of water vapor present, we have high humidity. Humidity has significant effect on how comfortable you are because, although your body may try to cool you by sweating, the air can't absorb much water so the moisture does not evaporate. Heat energy is required when water evaporates, so sweating can remove heat from your body as long as the air can absorb more water vapor. That is why people are more comfortable in "dry heat" than when the humidity is high.

When water vapor **condenses** it becomes liquid water through a phase change. Condensed water from the air causes dew, fog, and clouds that can result in rain or snow. Condensation is the opposite of evaporation—gas to liquid vs. liquid to gas. Heat energy is involved during phase changes such as when water evaporates or condenses. Energy exchanges are important in many aspects of our weather.

Humidity is normally reported as two different measures—**relative humidity** and **dew-point temperature**. Both measures tell us how wet or dry the air is. A hygrometer is used to measure humidity by measuring these heat-energy interactions.

This activity has two parts. You will measure relative humidity in Part 1 and study heat energy in Part 2.

OBJECTIVE: Understand the relationship between temperature, humidity, and heat energy.

Weather Tote:

Part 1:

- A hygrometer

  Instructions to make and use this hygrometer are in the Making Weather Instruments activity in Level 2. You may use a purchased hygrometer. If you do, follow the instructions to determine the relative humidity.

- A thermometer, similar to the hygrometer thermometers
- Relative Humidity and Dew-Point. Temperature Chart on page 76.

Part 2:

- A saucepan that holds at least 8 cups (2 qt.) to heat on a stove
- A high-range thermometer, one that will record a temperature greater than 212°F or 100°C (a lab thermometer or a candy thermometer)

Part 1:

1 Label your thermometers:

Thermometer 1 is called the dry-bulb thermometer.

Thermometer 2, connected to a jar of water, is called the wet-bulb thermometer.

Thermometer 3 will be used to determine the water temperature in the jar.

2 Prediction: Estimate the temperatures that you expect for each of the following:

Dry-bulb temperature (#1): ________

Wet-bulb temperature (#2): ________

Water temperature (#3): ___________

Relative humidity:________________

3 Take the lid off the jar of the hygrometer and leave the shoestring in the jar of water.

4 Place Thermometer 3 in the jar of water to determine the temperature of the water.

5 Wait five minutes and record the temperatures on the three thermometers:

Dry-bulb temperature (#1): ________

Wet-bulb temperature (#2): ________

Water temperature (#3): ___________

6 Calculate the relative humidity and dew-point temperature using the Relative Humidity and Dew-Point Temperature Chart by following these directions:

Find the dry-bulb temperature (column 1) with your left hand and the temperature difference (row 1) with your right hand.

Move your index fingers (left index finger moves to the right and right index finger moves down) to find the box in the chart where they intersect.

Read the two numbers in the intersecting box.

- The top number (black) gives the relative humidity.
- The bottom number (red) gives the dew-point temperature.

Relative humidity: ______________

Dew point: ____________________

Notes:

- If Thermometers 1 and 2 have the same temperature, the relative humidity is 100% and the dew-point temperature is the same temperature as your thermometers.
- If the wet-bulb temperature is higher than the dry-bulb temperature, your thermometers may not be calibrated correctly or the temperature has dropped rapidly outside. If this happens, wait an hour and take your reading again.

7 How did the values that you measured compare to the values you predicted?

8 Use your hygrometer in another situation (examples, below) to compare relative humidity levels.

A bathroom before and after a shower

An air-conditioned room compared to outdoors

On a morning with heavy fog or dew compared to later on the same day

At a swim meet, indoor pool compared to outside

Indoors when the heat is on compared to outdoors

9 Where did you measure the relative humidity? How did the values compare?

Temperature and Humidity

The temperature difference between the hygrometer's dry-bulb and wet-bulb thermometers allows you to calculate the humidity. When the air is humid, the difference in temperature is slight. As the air becomes dryer, evaporation of water on the wick increases and the difference between the dry-bulb and wet-bulb thermometers increases.

During the time that water is changing state (phase change), the temperature does not change, even though energy is still being added or removed. This absorption or release of energy is referred to as latent heat. Evaporation has a cooling effect as heat energy is absorbed. Condensation has a warming effect as heat energy is released.

Part 2:

Study heat energy:

1 Put 3 cups of water in a medium-sized saucepan.

2 Put the pan on the stove and bring the water to a boil.

3 Record the temperature of the water every minute from the time you begin until three minutes after the water comes to a rolling boil.

Use the high-range thermometer.

When recording the temperature:

- Hold the thermometer in the water so the bulb of the thermometer does not touch the pan.
- Read the temperature when the mercury stabilizes.
- Remove the thermometer from the water.

4 Record the temperatures below or make your own data sheet.

5 Make a graph of the change in temperature over time (x-axis: time; y-axis: temperature).

6 Note when the water began to boil in your temperature table and graph.

Be careful when you boil water. Boiling water can splash out of a pan and burn your skin, especially if the pan is too small. Make sure the pan has a secure handle and the capacity to hold at least 8 cups of water.

TEMPERATURE READINGS:

Time minutes	Temperature °F or °C	Time minutes	Temperature °F or °C
0			

QUESTIONS:

What energy is used to raise the temperature of the water?

What happened to the temperature readings when the water started boiling?

Why?

SHARE WHAT HAPPENED:

Part 1: Was the wet-bulb or dry-bulb temperature higher?

Part 2: What happened to the temperature once the water began to boil?

APPLY: Why do you think the wet-bulb temperature is generally lower than the dry-bulb temperature?

Will water evaporate slower or more quickly when air is very dry?

When your body begins to overheat, you perspire. Why does perspiration help cool your body?

What happens to perspiration on a hot, humid day?

In desert towns, people often talk about the weather having a "dry heat." Why would low humidity make hot weather more comfortable?

Why do people complain, "It's not the heat. It's the humidity?"

GENERALIZE TO YOUR LIFE: When you go swimming on a hot day your body has to to adjust to the cooler temperature of the water, and then the water feels fine. Why do you feel colder when you get out of the water and into the hot air?

If you are cooking something in boiling water on the stovetop, can you get the water hotter by increasing the burner's heat setting?

NOTES

HURRICANES

How does tracking the path and intensity of a hurricane show how hurricanes strengthen and weaken?

intro Hurricane Sandy made landfall on the New Jersey shore on October 29, 2012. Hurricane Katrina made landfall on August 29, 2005, on the Louisiana coastline in the Gulf of Mexico, just east of New Orleans. On August 24, 1992, Hurricane Andrew made landfall in southern Florida—first, Elliot Key in the Florida Keys, and then west to Homestead, a suburb of Miami.

The tremendous damage these massive storms caused rank them as three of the four costliest hurricanes in U.S. history. Scientists at the National Hurricane Center correctly predicted the track and intensity of these hurricanes and were able to give accurate warnings that likely saved many thousands of lives.

In this activity, you will track these three hurricanes and learn why they strengthened or weakened. All three storms started as tropical depressions far out in the Atlantic Ocean close to the equator. You will plot their courses to see how they turned from tropical storms into hurricanes.

OBJECTIVE: Track the paths of serious storms and analyze how they became hurricanes.

Weather Tote:

- A set of colored pencils or pens
- A blank copy of the Atlantic Basin Hurricane Tracking Chart on page 108

Do it

1 Use the hurricane data for the days given in Tables 10–12 to plot the path of each hurricane on your Tracking Chart.

Use a different color pencil or pen for each of the three hurricanes.

Mark each point with the day (1, 2, 3...), found in the first column of each table.

Mark each point with the wind speed, found in column 5 of each table.

Mark each point with the air pressure, found in column 6 of each table.

Create a key for your chart, showing the color you used for Andrew, Katrina, and Sandy.

Place your key in the bottom right corner of the chart.

2 Draw a line connecting the hurricane's location each day to mark each hurricane's path. Label the path of each hurricane close to the state it affected.

3 Use the following storm classifications to indicate the classification of each hurricane in the tables and your Tracking Chart.

Tropical depression:
maximum wind speed < 39 mph

Tropical storm:
maximum wind speed 39–73 mph

Category 1 hurricane:
maximum wind speed 74–95 mph

Category 2 hurricane:
maximum wind speed 96–110 mph

Category 3 hurricane:
maximum wind speed 111–130 mph

Category 4 hurricane:
maximum wind speed 131–155 mph

Category 5 hurricane:
maximum wind speed > 155 mph

TABLE 10. **HURRICANE SANDY | OCTOBER 21–31, 2012** — **CATEGORY:**

Day	Date in October 2012	Latitude	Longitude	Wind Speed (MPH)	Air Pressure (millibars)
1	22	13 N	79 W	40	1000
2	23	14 N	78 W	50	993
3	24	18 N	77 W	85	972
4	25	23 N	75 W	115	963
5	26	27 N	77 W	75	971
6	27	30 N	76 W	85	960
7	28	33 N	72 W	75	952
8	29	38 N	73 W	90	940
9	30	40 N	79 W	45	986
10	31	42 N	81 W	35	995

TABLE 11. **HURRICANE KATRINA | AUGUST 23–31, 2005** — **CATEGORY:**

Day	Date in August 2005	Latitude	Longitude	Wind Speed (MPH)	Air Pressure (millibars)
1	24	25 N	77 W	45	1003
2	25	26 N	80 W	70	988
3	26	25 N	83 W	95	968
4	27	25 N	85 W	115	948
5	28	26 N	89 W	174	902
6	29	31 N	90 W	90	948
7	30	37 N	87 W	35	990
8	31	40 N	83 W	30	996

TABLE 12. **HURRICANE ANDREW | AUGUST 16–28, 1992**				**CATEGORY:**	
Day	Date in August 1992	Latitude	Longitude	Wind Speed (MPH)	Air Pressure (millibars)
1	16	11 N	36 W	30	1010
2	17	13 N	44 W	35	1003
3	18	14 N	52 W	50	1000
4	19	19 N	58 W	50	1007
5	20	23 N	62 W	45	1014
6	21	25 N	65 W	55	1004
7	22	26 N	70 W	110	969
8	23	25 N	76 W	175	922
9	24	26 N	83 W	130	947
10	25	28 N	90 W	145	941
11	26	31 N	92 W	55	991
12	27	34 N	88 W	30	999
13	28	35 N	84 W	25	1000

Chat

SHARE WHAT HAPPENED:

What happens to the wind speeds as the air pressure falls?

What kind of surface was each hurricane over when it reached its highest wind speeds?

What category did each hurricane reach?

What happened to each storm's intensity as it made landfall? Why do you think this occurred?

What is the major difference between the paths of Hurricane Sandy and the other two hurricanes?

How do you think hurricanes develop when they typically start as strong storms?

APPLY: Figure 13 shows the tropical cyclone formation regions.

What two continents do hurricanes strike the most?

For the most part, at what latitude do hurricanes appear to start? Note: Compare Figure 13 to your tracking chart.

If you want to avoid hurricanes, should you live on the East Coast or the West Coast of the United States?

GENERALIZE TO YOUR LIFE: Do you know anyone who lives along the Gulf of Mexico or the East Coast of the United States? What should be part of a plan to prepare for a hurricane in these areas?

NOAA defines a tropical cyclone as a rotating, organized system of clouds and thunderstorms that originates over tropical or subtropical waters and has a closed low-level circulation. Classifications:

Tropical Depression: A tropical cyclone with maximum sustained winds of 38 mph (33 knots) or less.

Tropical Storm: A tropical cyclone with maximum sustained winds of 39 to 73 mph (34 to 63 knots).

Hurricane: A tropical cyclone with maximum sustained winds of 74 mph (64 knots) or higher. In the western North Pacific, hurricanes are called typhoons; similar storms in the Indian Ocean and South Pacific Ocean are called cyclones.

Major Hurricane: A tropical cyclone with maximum sustained winds of 111 mph (96 knots) or higher, corresponding to a Category 3, 4, or 5 on the Saffir-Simpson Hurricane Wind Scale.

Source: www.nhc.noaa.gov/climo

Fly Higher:

NOAA, The U.S. National Oceanic and Atmospheric Administration, has a lot of information. Visit any of the following websites and review the information available:

- National Hurricane Center, www.nhc.noaa.gov/
- Tropical Cyclone Climatology, www.nhc.noaa.gov/climo/
- Weather Safety webpage, www.weather.gov/safety/
- Hurricane Preparedness webpage, www.nhc.noaa.gov/prepare/

FIGURE 13. **TROPICAL CYCLONE FORMATION REGIONS**

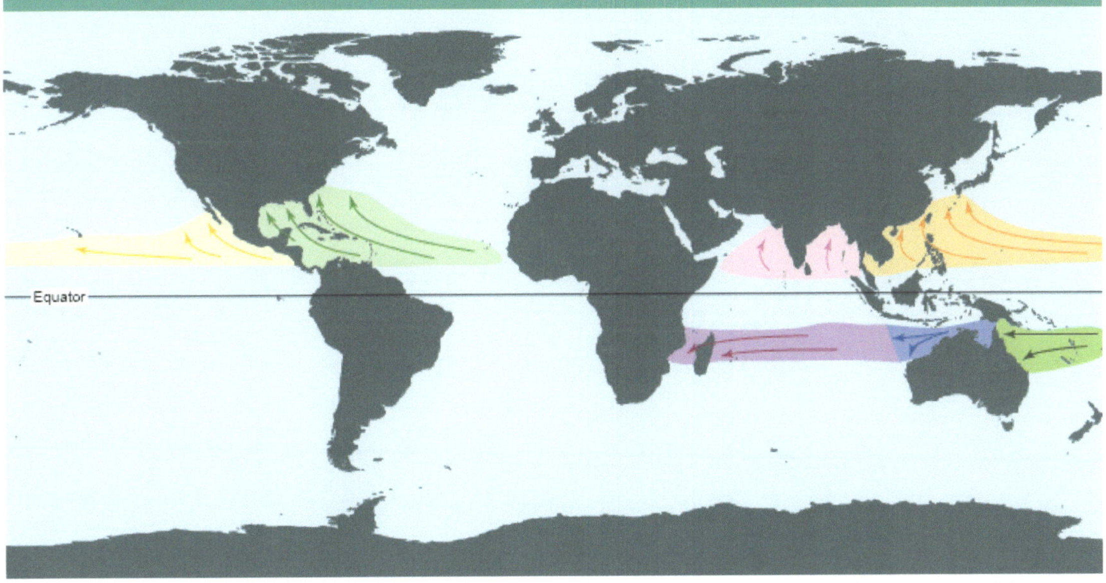

Source: National Oceanic and Atmospheric Administration/Department of Commerce

MAKING WEATHER INSTRUMENTS

Can I make my own weather instruments?

Yes, you can make your own weather instruments! While you can also purchase instruments to keep weather data, it is both fun and educational to make your own. You will learn how to make one or more of the following instruments:

1 Rain gauge

2 Wind vane

3 Anemometer

4 Barometer

5 Hygrometer

Each instrument is presented individually. The materials needed to make the instrument are given in the Weather Tote. Instructions to make the instrument follow in Make It. Use It gives a brief description on setting up and using your weather instrument. You can learn how to collect and record weather data in the Using Weather Instruments to Collect Data activity.

OBJECTIVE: Make one or more weather instruments.

Safety First!

Many of the weather instruments are made using tools that can be dangerous—a saw, knife, hot glue gun and liquid solder. Be particularly careful not to cut yourself when cutting the sheet metal or removing the tin-can lids. The glue gun generates high heat. The liquid solder creates a strong bond instantly, so do not get any on your hands or it might bond your fingers. Ask a parent or other adult to help when you see a safety first icon.

RAIN GAUGE A rain gauge measures the amount of precipitation.

Weather Tote:

- Two identical large tin cans or coffee containers, about 7" high x 4⅛" in diameter
- A funnel, about 5" to 5½" in diameter. The largest part should be wider than the can, or rainwater will bypass the funnel.
- A piece of wood, about 5" x 5" x ¾"
- A piece of wood, about 4" x 4" x ½" or foam (4" x 4" x 1")
- Jigsaw and drill if you use wood; knife if you use foam
- A 2" x 15" strip of metal or plastic
- A large rubber band, or an 18" length of wire
- A thin, straight measuring stick, about ⅛" x 3/16" x 12"
- Two or three tin cans, each with no more than 2" in diameter
- Nails
- Glue
- Ruler
- Hot glue gun and glue stick, or a small tube of liquid solder (pronounced "sod-er" and used to join metal pieces)
- A wooden post, 1½" x 2" in its cross section and 3'–4' long
- Permanent marker

RAIN GAUGE
FIGURE 14

RAIN GAUGE FUNNEL
FIGURE 15

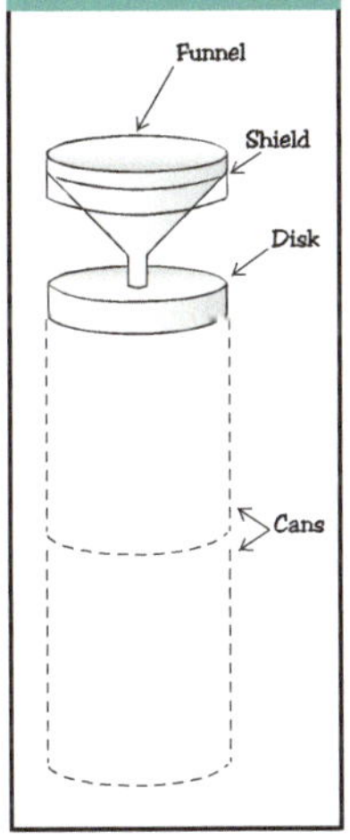

RAIN GAUGE HOLDER
FIGURE 16

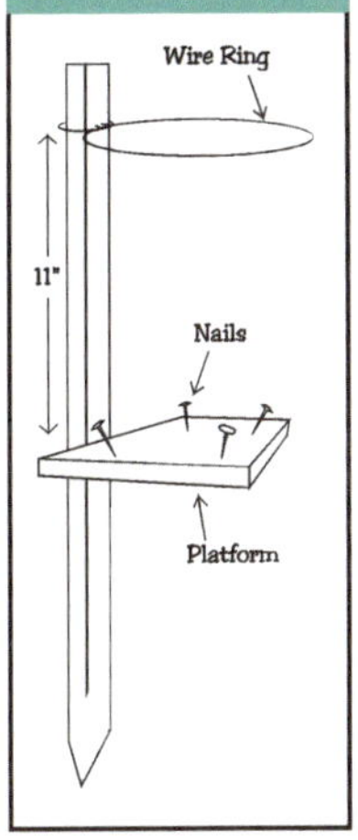

MAKE it Rain Gauge

1. Remove the tops from your containers.
2. Remove the bottom from one of the containers, and stack that can on top of the can with the bottom to make one 14" high can (see Figure 15).
3. Use the liquid solder or hot glue to bond the two cans together. **NOTE:** You only need to apply the liquid solder to one can.
4. Set the cans aside for 20–30 minutes to dry.
5. Cut the 4" x 4" piece of wood or plastic into a disc (circle) with a diameter about 3⅞".
   - Drill or cut a hole in the center of the disc, so the spout of the funnel just fits through.
   - Glue the wood or foam to the spout of the funnel.
   - This will weigh the funnel down, so it will not blow off or become tilted.
6. Apply the liquid solder or hot glue to the 2" x 15" piece of metal or plastic to the top of the funnel to form a shield to direct all rain through the funnel.
7. Place the funnel inside the top of the tall bonded can.
8. Mount your rain gauge on the post, as shown in Figure 16, by attaching the 5" x 5" x ¾" piece of wood to the post about 11" from the top. Hold it in place with 4 nails. Figure 16 shows how the nails are driven at an angle (toenailed) to hold the can.

TABLE 13. **CALCULATING RAINFALL**

RAIN GAUGE FORMULA	EXAMPLE CALCULATION
P = M (d/D)2	Using the rain gauge formula and M = 4"; d = 1"; and D = 5":
where:	P = 4 x (1/5)2 inches
P precipitation (inches)	P = 4 x (0.2)2 = 4 x (0.04) inches
M measured depth (inches)	P = 0.16"
d diameter of the smaller measuring can	The rainfall depth was approximately 2/10 inches (0.2), rounded to the nearest tenth of an inch.
D diameter of the funnel top	

9 Attach a wire ring to the post about 1" from the top.

Rain Gauge
Measuring Precipitation

1 Drive your rain gauge post into the ground in an open area away from trees and buildings that might affect your data.

- You can attach the post to a fence post if one is available in an open area.

2 Check your rain gauge shortly after a rainfall.

- Mark the rainfall depth on the outside of the container.
- Measure the rainfall depth in your rain gauge.

3 Use the rain gauge formula and instructions below to get a more precise rainfall reading.

- Remove the funnel from your rain gauge, and pour the water into another cylindrical can with a small diameter, preferably less than 2". NOTE: You can make a taller measuring can by cementing two cans together as you did for the rain gauge.
- Place the ruler in the can so it touches its bottom (like a dipstick); remove the measuring ruler and note the measurement, M, where the water wet the measuring stick.
- Use the rain gauge formula (Table 13) to calculate the amount of precipitation (P) that your area received.
- Indicate the amount of rainfall (P) on your container. Was P close to the amount that you measured?

4 Replace your rain gauge so it is ready for the next rainfall.

NOTE: Using the smaller diameter cylinder and rain gauge formula allows a more precise rainfall reading using multiplication. A smaller diameter cylinder is not used to collect rain because wind, which often accompanies rainfall, can cause too little rain to be collected for an accurate reading.

Fly Higher

Study how the diameter of a rain gauge affects your rainfall measurement by making three rain gauges with different diameters. You might try using your 2" measurement cylinder, your 5" diameter gauge, and a 5-gallon bucket or pail.

WIND VANE A wind vane measures the direction that the wind is blowing.

Weather Tote:

- Pieces of metal

  4" x 15"

  4½" x 5"

  ½" x 12"

  You can cut these metal pieces from a tin can.

- One ½" x 18" dowel, or a 1" x 1" x 18" stick
- Waterproof glue
- Nails, small
- Liquid solder or hot glue gun and glue stick
- Glass bead or washer
- Dowel or stick for mounting your vane

MAKE it Wind Vane

1 **VANE TAIL:** Fold the 4" x 15" piece of metal in half. See Figure 17 to see how to cut the folded end with a concave cut, so that the center 1.5" of the fold remains.

2 **ARROW POINT:** Fold the 4½" x 5" metal to form a triangle with sides 4½" x 2½" (see Figure 18).

3 Cut parallel 1" slots on each end of the dowel rod, about 1" long and ½" thick. Be sure that the slots are parallel (lined up).

4 Attach the arrow point and vane tail using small nails and waterproof glue.

5 **TAIL SEPARATOR:** Fold the ends of the 12" x ½" piece of metal to form a V that is 2.5" high. See Figure 20 to see how to solder or hot-glue the tail separator.

6 Place the shaft (dowel) on the edge of a knife. This is called the balancing point.

7 Drill a small hole through the shaft at the balancing point and parallel to the folded edge of the vane.

- If the balancing point is too far back, shift it forward by adding weight to the arrowhead end by wrapping heavy wire or a similar material to the front of the shaft.
- If the balancing point is too far forward, shift it backward by adding weight to the arrowhead end by wrapping heavy wire or a similar material to the tail end of the shaft.

8 Mount the wind vane on an upright post or stick with a nail that fits loosely through the small hole that was drilled at the balancing point.

9 You may attach a glass bead or washer between the vane shaft and the post as a bearing to make the vane move more freely. See Figures 20 and 21 for assembly details.

FIGURE 17. **VANE TAIL**

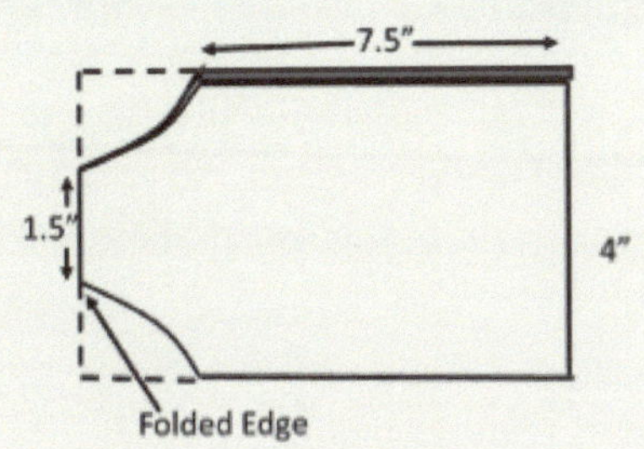

FIGURE 18. **ARROW POINT**

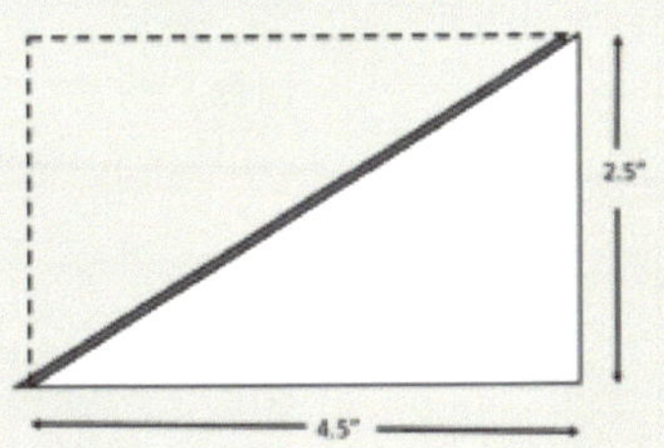

FIGURE 19. **TAIL SEPARATOR**

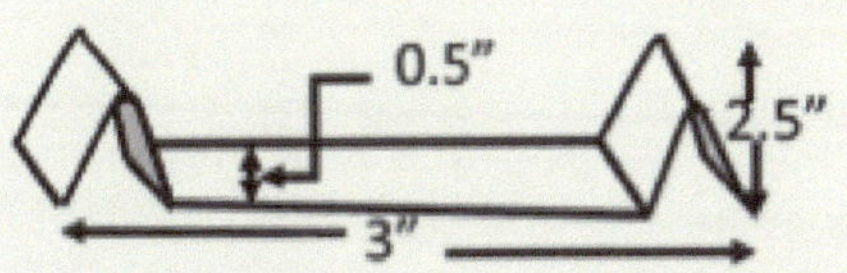

Wind Vane
Determining Wind Direction

1 Mount or place your wind vane in an open area. Try to avoid obstructions such as trees or buildings that can affect the wind.

2 The arrowhead will point in the direction that the wind is blowing: north, south, east, west, or in between.

FIGURE 20. **WEATHER VANE CONSTRUCTION**

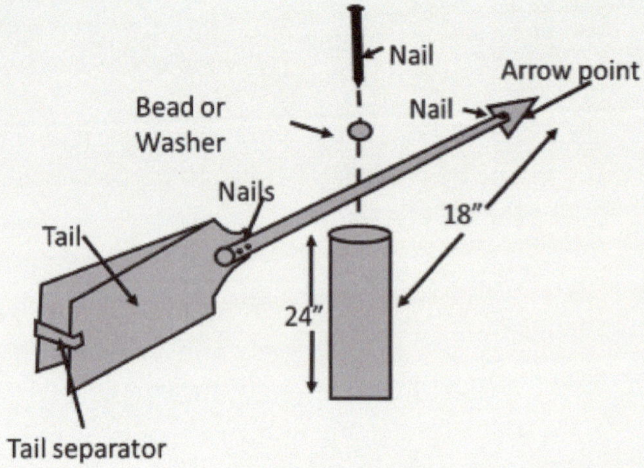

FIGURE 21. **WEATHER VANE**

ANEMOMETER Wind speed is often measured by a cup anemometer. Cup anemometers can be made from funnels, tin cans, or plastic cups.

Weather Tote:

- Three metal funnels, approximately 2½" in diameter (See Figures 22 and 23.)
  - Or: three pieces of sheet metal, approximately 2½" x 5"
  - Or: three plastic cups, approximately 2½" in diameter
- Dowel, 3/16"x 17"
- A metal or plastic pencil or pen cap
- A metal lid with 1¾"–2" diameter
- A small washer with ⅜" diameter, or a piece of tin
- A 3" finishing nail
- Paint
- Waterproof glue
- Liquid solder or hot glue gun and glue stick
- Commercial anemometer (borrowed) or radio, TV, or online weather source for daily wind-speed report

Anemometer

NOTE: An anemometer made with plastic cups will be fairly sensitive to light winds, so it may be difficult to count the revolutions if the wind is blowing very hard. One made of flat metal or metal funnels will turn more slowly and therefore may measure higher winds more easily.

1 Cut and shape the spout, if you are using metal funnels. See Figure 22.

- Cut the spout, leaving only about ½”.
- Cut three V-shaped slits in the ½” spout, and bend the pieces toward the neck of the funnel.

2 Solder or hot glue the opening shut to keep wind from blowing through.

3 Punch a hole on opposite sides of your three funnels so that the 3/16” dowel rods fit tightly through, as shown in Figures 22 and 23. Fasten each with waterproof glue.

4 Use a small bottle lid to assemble the hub of the anemometer by drilling or punching three equally spaced holes in the rim, as shown in Figure 24.

5 Drill or punch a hole in the center of the lid, and glue a ballpoint pen cap or other plastic cap in the hole.

6 Glue the washer to the opening of the pen cap.

7 Insert the funnel-and-dowel spokes into the lid so they touch the pen cap and rest on top of the washer. See Figures 24 and 25.

8 Check that the funnel is parallel to the axis of the pen cap and all point in the same direction.

9 Glue the spokes to the cap and the washer.

10 Paint one funnel so that it is easy to see (and count) when your anemometer is spinning.

11 Mount the anemometer to the wood post by pounding a 3” finishing nail into the post and placing the anemometer on top of the nail.

12 Check to see that your anemometer turns freely.

- If it does not, try cutting off the head of the nail and sharpening it to a point with a file.
- Lubricating this point with oil may help the anemometer move more easily.

HANDMADE ANEMOMETER

FIGURE 22. FUNNEL WIND CATCHER

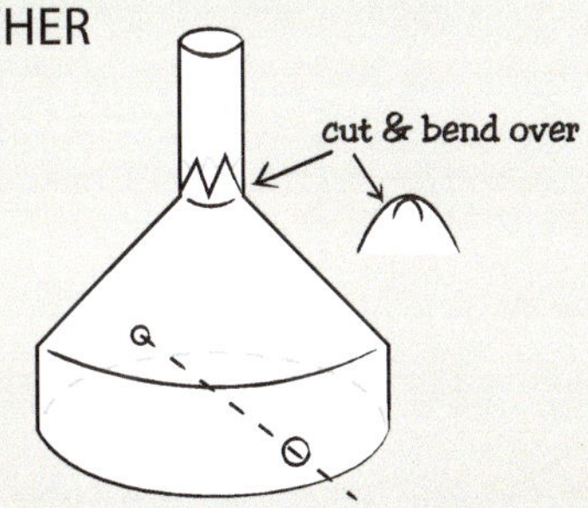

FIGURE 23. SHEET METAL WIND CATCHER

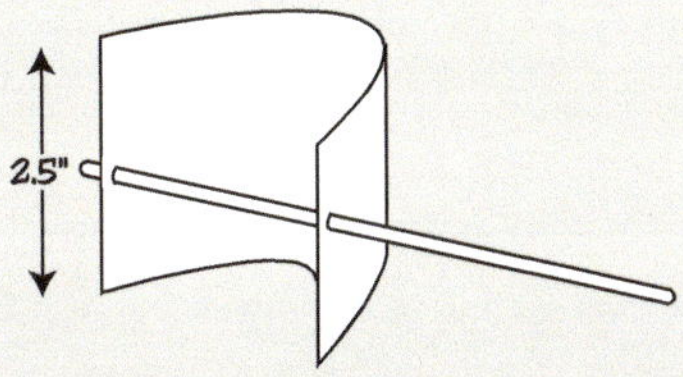

FIGURE 24. ATTACH BASE TO TOP

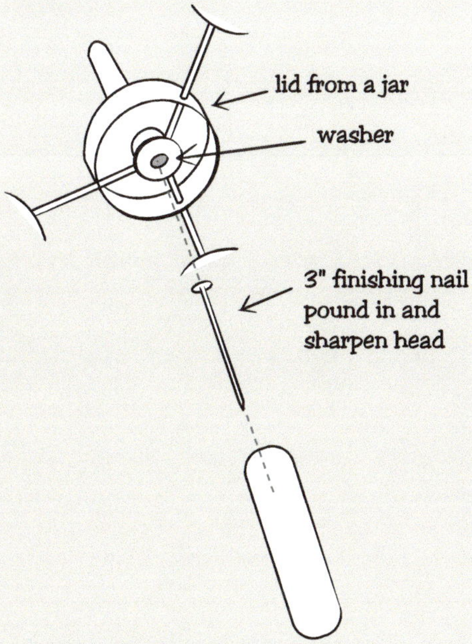

FIGURE 25. FIGTOP WITH CUPS/FUNNELS

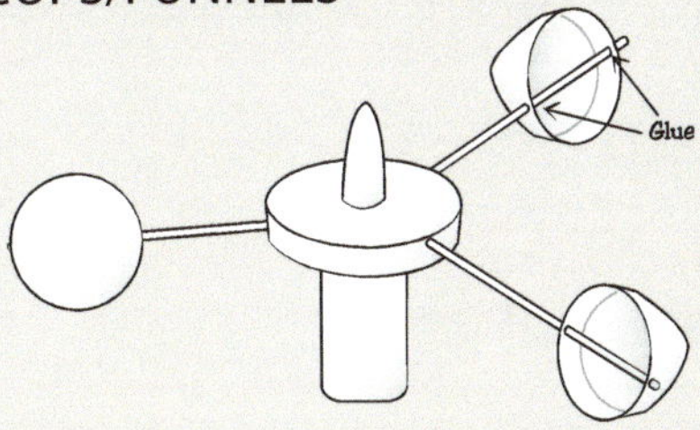

FIGURE 26

Anemometer
Determining Wind Speed

1 Mount or place your anemometer in an open area. Try to avoid obstructions such as trees or buildings that can affect the wind.

2 Before using your anemometer you must calibrate or calculate the wind speed in miles per hour, mph, as described below.

3 Count the number of revolutions in one minute.

4 Record the date, time, and wind speed.

TABLE 14. **WIND SPEED DATA CHART**

Reading	Date	Time	Revolutions Per Minute (RPM) x-axis	Miles Per Hour (MPH) y-axis
1				
2				
3				
4				
5				
6				

WIND SPEED GRAPH

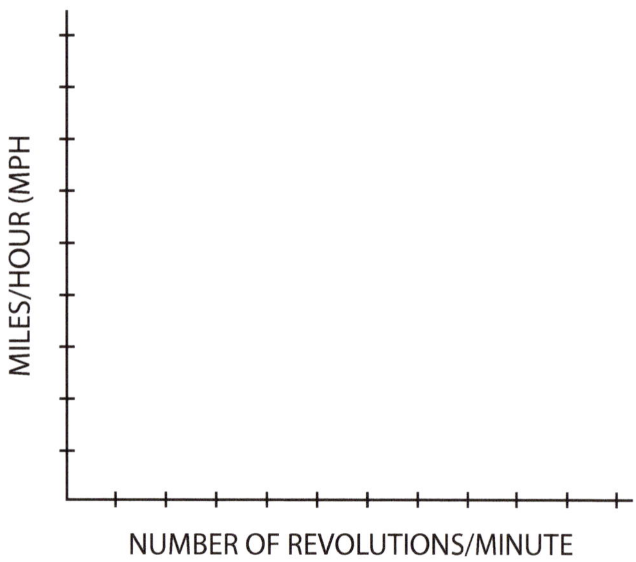

DETERMINING THE WIND SPEED

Wind speed is reported in miles per hour, mph, so you need to convert the number of revolutions of your cup anemometer to wind speed in mph. Choose one of the following three methods to determine wind speed from your anemometer: anemometer calibration; calculation; or vehicle calibration. Calibration is a process of comparing an unknown value to a known value. The three methods are described below.

Anemometer Calibration

This method uses a second anemometer for calibration. You can purchase or borrow an anemometer or use the wind speed that is reported in the media from a source near you. Choose how you would like to calibrate your anemometer (A or B) and follow the steps to create a Wind Speed Data Chart for your anemometer.

A Use a purchased or borrowed anemometer that records the wind speed in mph.

- Count the number of revolutions for one minute (RPM) on your handmade anemometer.
- Note the purchased anemometer's wind-speed reading in mph.
- Record your data in the Wind Speed Data Chart following the directions for recording data (page 88).
- Collect 10 data points.
- Make a Wind Speed Graph following the directions for recording data (page 88).

B Check a local weather source on radio or TV or online for the reported current wind speed. Using a weather report will not be as accurate as reading from another anemometer in your location because the weather station that is reporting may be miles away from your anemometer, and winds can vary from place to place.

- Count the number of revolutions for one minute (RPM) on your handmade anemometer.
- Obtain the wind speed reported in mph by your weather source.
- Record your data in the Wind Speed Data Chart following the directions for recording data.
- Collect 10 data points at least 4 hours apart.
- Make a Wind Speed Graph following the directions for recording data (page 9).

Calculation

Use mathematics to calculate the wind speed.

1 Measure the radius of the circle described by the funnels—the distance from the center of your anemometer to the center of one of the funnels, cups, or flat metal pieces. They should each measure the same distance.

2 Count the number of anemometer revolutions for one minute.

3 Calculate the wind speed using the following formula:

0.00595 x the radius of the anemometer in inches x number of revolutions per minute

This formula was determined by multiplying the distance traveled by the funnels or cups ($2\pi r$, the circumference, or distance traveled) by 60 (minutes in an hour); dividing this value by 12 (inches in a foot); and dividing that by 5,389 (feet in a mile), which will convert the rpm to mph. The radius, r, is measured in inches.

4 Record your data in the Wind Speed Data Chart following the directions for recording data (page 88).

5 Collect 10 data points at least 4 hours apart.

6 Make a Wind Speed Graph following the directions for recording data (page 88).

Vehicle Calibration

Use an automobile's speedometer (mph) to calibrate your anemometer. This method is the last choice because of safety considerations. It can be used, however, if you enlist the help of a careful, competent adult driver and have access to a straight stretch of a deserted country road.

1 Mount your anemometer securely on the handle of a broomstick or a similar rod.

2 Ask the adult to drive at a constant speed for one minute at each speed—5, 10, 15, and 20 mph.

3 Hold your anemometer out of the car window, well above the car roof, and count the number of revolutions at each speed. Note: If the wind is blowing when you are calibrating your anemometer, you must repeat the procedure at the same speed in the reverse direction. The average of the two measurements will give you a more accurate calibration.

4 Record your data in the Wind Speed Data Chart following the directions for recording data.

5 Collect 10 data points at least 4 hours apart.

6 Make a Wind Speed Graph following the directions for recording data.

Recording Data

Record data in the Wind Speed Data Chart and make a Wind Speed Graph following the directions below. Then you can use your Wind Speed Graph to find the wind speed in mph based on the number of revolutions of your cup anemometer.

1 Take at least 10 readings at least 4 hours apart.

2 Record the following data on the Wind Speed Data Chart.

- Date, time, and RPM (from your handmade anemometer)
- Wind speed from the anemometer or media report

3 Graph the 10 data points (RPM, MPH) on the Wind Speed Graph.

- Point to the RPM on the x-axis and the MPH on the y-axis and draw an imaginary straight line from each until they intersect. Place a data point there.

4 Connect the data points and extend the line.

5 Use the Wind Speed Graph to determine the wind speed:

- Count your anemometer's RPM.
- Find that value on your graph.
- Follow an imaginary vertical line up to your data line.
- Follow an imaginary horizontal line to the y-axis.
- Read the wind speed in mph.

BAROMETER A barometer measures pressure changes but will respond to both pressure changes and temperature changes, so it must be kept at a constant temperature to measure pressure accurately. Barometers indicate if the air pressure is rising or falling.

Weather Tote:

NOTE: There are two methods for making a barometer using either a tin can or baby food jar. You may make either one.

- Relative Humidity and Dew-Point Temperature Chart on page 107
- Aneroid barometer or weather source for the report of the current barometric pressure
- 1" x 8" x ¾" piece of wood
- Glue or rubber cement
- Pointer made from a small stick, Popsicle stick, or small flattened drinking straw
- Index card
- Choose the type of barometer you want to make, and gather the additional items listed.
    - The baby food jar barometer is easiest to make.
    - Do not use plastic containers because they will give inaccurate readings.

TIN-CAN BAROMETER

- Tin can with two metal ends
- Wood, 6" x 15 x ¾"
- Wood, 1" x 8" x ½"
- Cork from a small bottle
- Thin stick or broom straw
- Toothpick
- Three screws
- Nails
- Liquid solder or hot glue gun and glue stick

Figure 27

BABY FOOD JAR BAROMETER

- Baby food jar
- Wood, 6" x 6" x ¾"
- Wood, 1" x 8" x ½"
- Piece of rubber—a balloon cut open will work
- Rubber band(s)

NOTE: Balloon rubber is thin and will slowly lose air, so this barometer will only be accurate for a couple of weeks. You will need to replace the balloon rubber every two weeks.

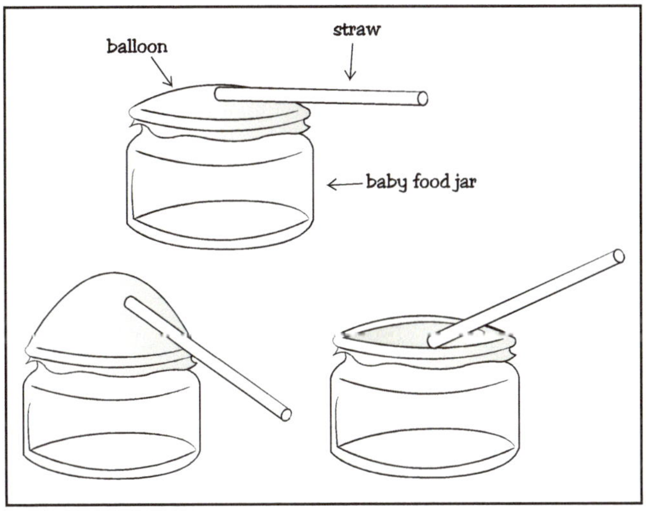

Figure 28

MAKE it Barometer

NOTE: You can use any size of can for this instrument, but a larger can (such as a coffee can) will give better readings. It is more sensitive because it has a wider surface for the air pressure to act on.

TIN-CAN BAROMETER

1. ✚ Nail the 1" x 8" piece of wood to the 6" x 15" baseboard. See Figure 27.
2. Attach the can to the baseboard with three equally spaced screws around the outside of the can. The screw heads will hold the can in place. Or glue a piece of foam, the size of the can, to the wood and place the can on top of it.
3. ✚ Attach the lid to the tin can using liquid solder or hot glue, leaving one small hole for air to escape.

4 Heat the can in hot or boiling water to force out as much air as possible.

5 Seal the hole immediately with liquid solder or hot glue.

6 Glue the cork on the top of the can just to the right of center.

7 Glue a thin stick or broom straw to the cork. Glue a toothpick to the end of the stick/broom straw. This is the "pointer."

8 Glue an index card to the piece of wood, positioned so the pointer is near the center of the card.

BABY FOOD JAR BAROMETER

1 Remove the lid from the jar. See Figure 28.

2 Stretch the piece of rubber over the opening, and attach it to the jar tightly with one or more rubber bands.

3 Nail the 1" x 8" piece of wood to the baseboard to form an upright stand.

4 Glue the jar to the 6" x 6" baseboard, so it won't move.

5 Glue the drinking straw to the top of the jar. This is the "pointer."

6 Glue the index card to the stand, positioned so the pointer is near the center of the card.

USE it

Barometer

Determining Air Pressure Trends

You can use your barometer to determine if the air pressure is rising or falling. But before you use your barometer you will need to calibrate it. Calibration is a process of comparing an unknown value to a known value (standard).

- Compare your barometer daily for a month by using a borrowed aneroid barometer or your radio, TV, or online weather source as your standard.
- Mark the index card where the pointer indicates and write the value.
- Note if the pressure is rising or falling.

NOTES:

- When the air pressure is low, the air inside the jar will expand and press on the rubber. This will cause the straw to point down.
- When the air pressure is high, the straw will depress the rubber and cause the straw to point up.
- Barometer changes are generally slight. They are most interesting when the temperature is rising or a storm is approaching.

HYGROMETER

A hygrometer is used for measuring the moisture content in the air.

You will use two thermometers to make a hygrometer that can be used to determine humidity, the amount of moisture in the air. Most scientists use a sling psychrometer, which also uses two thermometers, one of which is covered by wet cloth. It is twirled in a circular pattern. The dry psychrometer determines the temperature of evaporation more quickly than the hygrometer in this activity. Both methods use the Relative Humidity and Dew-Point Temperature Chart to determine the relative humidity and dew-point temperature.

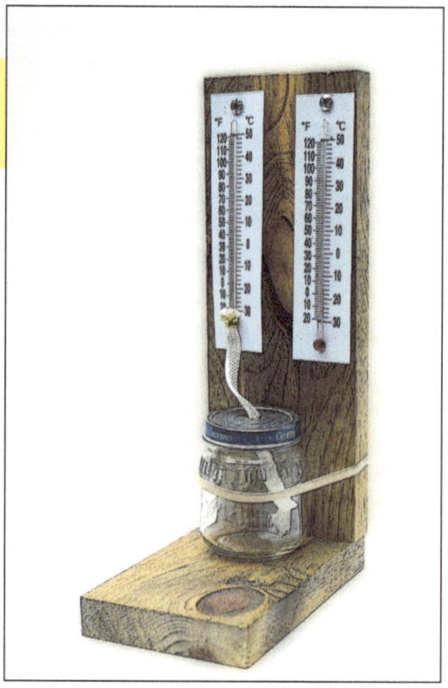

FIGURE 29.
HYGROMETER

Weather Tote:

- Wood, 1" x 6" and about 2 feet long
- Piece of hollow shoestring, about 6" long
- Scrap wood, about 2" x 4," for drilling
- Saw
- Hammer
- Nails (1⅝" panel nails work well)
- Jar with lid—a baby food jar or similar glass or plastic jar
- Two similar thermometers, purchased or borrowed
- A long twist tie
- Scissors
- Drill and ¼" drill bit
- Relative Humidity and Dew-Point Temperature Chart (page 93)

MAKE it Hygrometer

1 Cut the wood to form a stand (see Figure 29).

- The baseboard should be about 6," large enough to hold your jar and to give the stand stability.
- The upright board needs to be tall enough to hold the thermometers about an inch or two above the jar placed on the baseboard.

2 Set the jar on the wooden stand.

3 Mount the thermometers side by side on the upright board, so they are an inch or two above the top of the jar.

4 Compare the temperature readings on the two thermometers.

- If they do not record the same temperature, you may be able to adjust one thermometer.
- If you cannot adjust the temperature on your second thermometer (Thermometer 2), note the difference in temperature and add or subtract that difference for all your readings for this thermometer.

5 Drill a ¼" hole in the center of the jar lid, with the lid held on a block of wood so that you drill through the lid and into the wood.

6 Place the hollow shoestring over the bulb of Thermometer 2.

- Secure the shoestring with a twist tie above the bulb of the thermometer.
- Cut off the excess twist tie with scissors.

7 Fill the jar with water.

8 Push the shoestring through the hole in the jar lid, so it lays on the bottom of the jar.

9 Attach the lid to the jar.

10 Secure the bottle to the stand with a rubber band.

- Push any extra shoestring into the jar so the shoestring above the lid does not sag.
- Make sure the shoestring is drawing water to Thermometer 2.

USE it Hygrometer

Measuring Temperature and Relative Humidity

1 Place the hygrometer outdoors in the shade, away from the sun.

2 Make your own table with 8 rows and 5 columns. You can use a piece of paper or computer.

- Row 1, column headers: Dry-Bulb Temperature (°F), Wet-Bulb Temperature (°F), Temperature Difference (°F), Relative Humidity (%), and Dew-Point Temperature (°F)
- Rows 2–8, columns: the dates for collecting this data for a week, starting with the current date.

3 Read the dry-bulb temperature and wet-bulb temperature from your hygrometer.

4 Record data each day for a week.

- Thermometer 1 records the dry-bulb temperature and gives the air temperature.
- Thermometer 2 records the wet-bulb temperature. Remember to correct the temperature on thermometer 2, if needed.
- Calculate the temperature difference between your readings of Thermometer 1 and thermometer 2.
- Follow these directions to find the relative humidity and dew-point temperature using the Relative Humidity and Dew-Point Temperature chart.

  Find the dry-bulb temperature (column 1) with your left hand and the temperature difference (row 1) with your right hand.

  Move your index fingers (left index finger moves to the right and right index finger moves down) to find the box in the chart where they intersect.

  Read the value of the relative humidity and dew-point.

  - The top number (black) gives the relative humidity.
  - The bottom number (red) gives the dew-point temperature.

  EXAMPLE: If the dry-bulb temperature is 62°F and the wet-bulb temperature is 55°F, the difference is 62 minus 55=7.

  - Column 1: mark 62 with your left index finger.
  - Row 1: mark 7 with your right index finger.
  - Move your fingers (left to the right; right moves down) to find the intersecting box.
  - Read the relative humidity as 64 and the dew-point temperature as 49.
  - If the dry bulb and wet bulb have the same temperature, the relative humidity is 100%, and the dew-point temperature is the same temperature as your thermometers.
  - If the wet bulb is higher than the dry bulb, your thermometers may not be calibrated correctly or the temperature has dropped rapidly outside and it will take some time for the wet bulb to adjust. If this happens, wait an hour and take your reading again.

SHARE WHAT HAPPENED:

- What weather instrument(s) did you make?
- Which instrument was the most fun to make or use?
- What was most challenging?

APPLY: What did you learn about the weather while you were making your instruments?

GENERALIZE TO YOUR LIFE: Why do you think people are so interested in the weather?

Fly Higher:

Review and compare the instructions in the NOAA publication, *Build Your Own Weather Station*, with the instructions given in this activity on pages 72–80.

Purchase or make a sling psychrometer (instructions can be found online) and compare relative humidity readings with the hygrometer you made.

Collect pictures of different ornamental wind vanes.

MINI-TORNADO

How can a water vortex created in a bottle help you understand tornadoes?

intro Tornadoes have always been both fascinating and dangerous. Capable of destroying entire cities, tornadoes can produce the highest wind speeds on earth. A tornado striking near Moore, Oklahoma, on May 3, 1999, had a measured wind speed of 318 mph, the highest wind speed ever recorded.

Our understanding of weather has increased significantly during the past thirty years, allowing earlier warning of potential or imminent tornadoes. Nevertheless, questions remain about why some severe thunderstorms produce tornadoes and others do not.

This activity demonstrates how water, like air, can create a vortex. A vortex illustrates the forces that help create a tornado's visual structure.

OBJECTIVE: Generate a water vortex that simulates a tornado and observe its progression.

Weather Tote:

- Two clear 2-liter plastic soda bottles, empty and clean
- Tornado Tube plastic connector

Available from science museums, science stores, novelty stores, scientific supply companies, and teacher-materials shops

Or make your own with a metal washer, about 1 inch in diameter with a 3/8-inch diameter hole, and plastic electrical tape.

- Liquid dishwashing detergent
- Food coloring (optional)
- Stopwatch or clock with a second hand

1. Fill one soda bottle about two-thirds full with water.
2. Add a drop of dishwashing detergent and a couple drops of food coloring to make the vortex easier to see.
3. Screw the plastic connector onto the top of the bottle.
4. Attach the empty bottle to the open end of the connector or flat washer (between the bottles) and electrical tape.
5. Turn the two-bottle assembly over, and replace the assembly on the flat surface with the filled bottle on top.
6. Watch the water drip down into the lower bottle as air simultaneously bubbles up into the top bottle. The flow of water may even come to a complete stop.
7. Rapidly spin the bottles in a circle a few times, keeping the bottom bottle in contact with the flat surface.
8. Observe what happens in the bottles.

EXPERIMENT WITH YOUR VORTEX CREATOR:

- Change how you spin the bottles. What happens?
- Change how long you spin the bottles. What happens?
- Observe any differences in the shape and length of time the vortex lasts.

Almost 99 percent of tornadoes spin counter-clockwise in the Northern Hemisphere. Test both rotations (counterclockwise or clockwise) with a stopwatch to see which has a longer-lasting funnel. Was there a difference, or were they the same?

SHARE WHAT HAPPENED:

- What caused the funnel to form inside the bottle?
- How did the funnel change as time passed?
- Did the "tornado" continue until the water flowed out, or did it stop before the water ran out? Explain why you think this happened.
- Did experimenting with the initial force cause differences in the shape and length of time the vortex lasted? Explain your findings.

APPLY: Look at the photos of tornadoes below and answer the questions that follow.

NOAA - Source: National Oceanic and Atmospheric Administration/Department of Commerce.

- Which image is most like the vortex you observed in the bottle?
- What is different about the images?

GENERALIZE TO YOUR LIFE: If a tornado warning is issued for your area, where is the safest place to go in your school? Your home? While in a car or bus on the road?

OUT OF THE DUST

What was the Dust Bowl? How did it affect the people who lived through it?

intro The Dust Bowl is the name given to a period characterized by severe dust storms that occurred in the Great Plains of the United States and Canada. From 1930 to 1949—primarily 1930–1936—the prairie soil dried, turned to dust, and blew eastward and southward in large dark clouds.

Decades of extensive farming, without conservation to prevent erosion, caused the Dust Bowl. The Great Plains were very fertile during times of adequate rainfall. When farmers plowed and planted, they unknowingly dislodged the natural grasses that kept the soil in place and trapped moisture.

The Dust Bowl affected 100 million U.S. acres, especially in the panhandles of Texas and Oklahoma, and adjacent land in New Mexico, Colorado, and Kansas. Immense dust storms with names like "Black Blizzards" and "Black Rollers" reduced visibility to a few feet. At times storms deposited soil into the Atlantic Ocean, and clouds blackened the sky as far away as New York City and Washington, DC.

Dust storms caused major ecological and agricultural damage and occurred during this country's worst economic depression. Many people had to leave the farms that they had built. Some even died.

The novel *Out of the Dust*, by Karen Hesse, is a Newbery Medal–winning story about the Dust Bowl. It is a work of fiction based on actual events and actual people, and tells the story of a young girl who stayed in Oklahoma and persevered. Note: This book has very graphic images of injuries the young girl's family suffered after a kerosene-fueled fire.

Dust storm

OBJECTIVE: Understand that weather can change people's lives and fortunes.

Weather Tote:

- *Out of the Dust* by Karen Hesse. [Hesse, Karen. *Out of the Dust*. New York: Scholastic, 1999. (ISBN: 978-0-590-37125-4)]

DO it

1. Read *Out of the Dust* by Karen Hesse.
2. Discuss the Chat questions with your adult leader, parent, or guardian.

SHARE WHAT HAPPENED:

How did the drought and dust storms affect Billie Jo's day-to-day life?

In any life experience, factors like the weather are only partial causes for changes. What other circumstances contributed to the changes in Billie Jo's life?

What kind of student was Billie Jo, and what were her talents? In what ways were her talents like yours?

How were Billie Jo's teachers important?

What did Billie Jo learn when she tried to run away?

What did you think was the most interesting part of the book? Why?

APPLY: How did the Dust Bowl change agricultural practices? Are any of these practices used today?

GENERALIZE TO YOUR LIFE: Can you think of weather events—either positive or negative—that have changed things in your life or the lives of your family members? Explain your answer.

Fly Higher:

Read more about the Dust Bowl. For example, research what happened to the many Dust Bowl families who moved to California with the hope of finding a new life.

Learn more about the Dust Bowl by watching the PBS documentary, *The Dust Bowl*, by Ken Burns, www.pbs.org/kenburns/dustbowl/.

NOTES

SEASONS

What causes the changes in the yearly weather patterns we call seasons?

intro We are used to seeing many different kinds of weather in the Midwest. Although we might experience a thunderstorm in January or snow in May, we know that winters are usually cold and snowy, and summers, hot and humid. We anticipate the general weather pattern that is based on our climate. The weather changes as the seasons change, and nature responds to the changes in weather. For example, we expect the trees to grow leaves in the spring and to lose them in the fall.

The sun provides the energy that warms earth. Variations in the tilt of earth as it revolves around the sun are responsible for seasonal changes. Many people mistakenly assume that summers are warmer because earth is closer to the sun at that time of the year. Because earth's orbit is not a perfect circle—it is an **ellipse**—earth is actually closest to the sun in January and farthest from the sun in July! This 3 percent difference in distance is very slight and has little or no effect on our seasons.

The tilt of earth is responsible for seasonal changes. The axis of earth is tilted at 23.5° in relation to earth's orbit around the sun. You will explore how this tilt causes the seasons.

OBJECTIVE: Understand how changes in the earth-sun relationship cause seasons.

Weather Tote

- A table
- A standard world globe on a tilted stand. You may borrow one from your school or library.
- A light source, such as a lamp, without its shade
- A dark room
- A protractor to measure the tilt of the globe.

DO it Perform this activity in a dark room, using the lamp as the only light source.

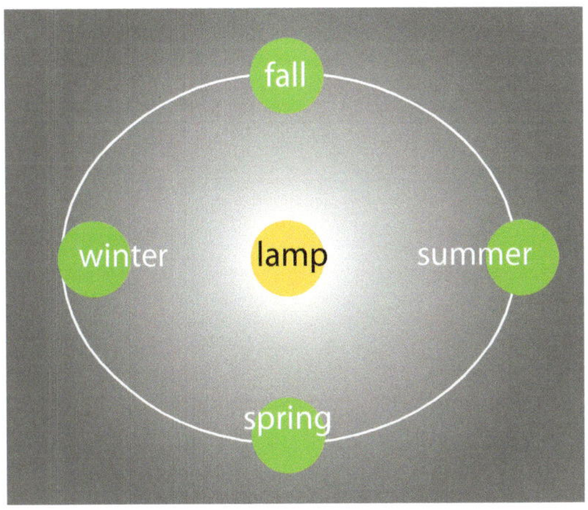

FIGURE 30. **EARTH'S ORBIT**

1. Place the lamp in the center of a table.
2. Set the globe on the right side of the table at the "summer" position shown in Figure 30.

3 Tilt the North Pole toward the lamp about 23 degrees.

The globe must be at the same height as the light bulb.

Raise the level of the globe by placing it on a stack of books, if necessary.

The resulting light shine represents the first day of summer, known as the summer **solstice**.

4 Turn the globe slowly counterclockwise to experience one day.

Find where you live on the globe. Observe what happens to the light there.

Observe how the light shines on both the North Pole and the South Pole.

5 Move the globe to the other three positions shown in the figure: Autumnal **Equinox** (fall), Winter Solstice (winter), and Vernal Equinox (spring).

Keep the tilt of the globe at 23 degrees, always leaning toward the left. This will track the actual progress of earth through the seasons.

8 Observe and record what happens to the light where you live and at both poles as you rotate the globe.

OBSERVATIONS:

Summer ______________________________

____

Fall ______________________________

____

Winter ______________________________

____

Spring ______________________________

____

SHARE WHAT HAPPENED:

- Do people in the Midwest receive more sunlight in the summer or winter?
- During the summer is it ever dark at the North Pole? If so, when and why?
- How might these differences affect our seasons?
- Contrast the difference between the summer and winter sunlight at the North Pole.
- Your globe might have the Tropic of Cancer marked at 23.5° north of the equator. If it does, notice that on the first day of summer, the sun stays directly over that line. What line does the sun stay over on the first day of winter?

APPLY: Calculate the total amount of daylight for each city:

FIRST DAY OF SUMMER:

Summer Solstice	Sunrise	Sunset	Daylight Length
Indianapolis, IN	6:17 AM	9:21 PM	
Miami, FL	6:30 AM	8:15 PM	
Anchorage, AK	4:20 AM	11:42 PM	

Note: Summer times are daylight saving time

FIRST DAY OF FALL:

Autumal Equinox	Sunrise	Sunset	Daylight Length
Indianapolis, IN	7:33 AM	7:40 PM	
Miami, FL	7:10 AM	7:15 PM	
Anchorage, AK	7:47 AM	7:55 PM	

Note: Daylight saving time

FIRST DAY OF WINTER:

Winter Solstice	Sunrise	Sunset	Daylight Length
Indianapolis, IN	8:06 AM	5:23 PM	
Miami, FL	7:03 AM	5:35 PM	
Anchorage, AK	10:14 AM	3:42 PM	

Note: Winter times are standard times

- Using your globe and lamp, explain why daylight lengths are different at these three locations at the summer solstice and the winter solstice.
- Using your globe and lamp, explain why daylight lengths are similar at these three locations at the autumnal equinox.
- What would be the daylight length on the first day of spring (vernal equinox)?
- We have seasonal changes in the Midwest. Do you think seasonal changes are more or less obvious in Alaska? Explain your answer.
- Why do you think Miami, Florida, has little seasonal change?

GENERALIZE TO YOUR LIFE: The actual **scale** of the earth and sun cannot be shown in diagrams used to explain the seasons. This is because the distances that are involved are so large. The sun has a diameter of 865,000 miles, which is more than 100 times the diameter of earth (7,900 miles). If the sun was the size of a basketball, earth would be about the size of a BB pellet, and the moon would be the size of a speck of dust. Earth is also in orbit 93,000,000 miles away from the sun. That would put the BB that represents earth 74 feet away from the basketball!

- Where could you use the actual earth/sun scale to help people visualize how far we are from the sun? Sketch it.
- How might you use scale in planning the layout of your bedroom when adding a new dresser or larger bed?

Fly Higher

- Compare the actual location of the rising and setting sun at different times of the year from your house—especially on the first day of each season.
- Record the length of a shadow at about noon during different seasons throughout the year. Use an existing pole or the shadow your house makes.

NOTES

USING WEATHER INSTRUMENTS TO COLLECT DATA

What can your weather instruments tell you?

intro

Weather observers record weather conditions all over the world. Some people have kept records for many years. They may keep data on temperature, **precipitation**, wind speed and direction, cloud conditions and types, **air pressure**, humidity levels, and when conditions create fog, haze, and smoke. In this activity you will join other weather watchers when you use your weather instruments to keep daily records of one or more of the following: precipitation, temperature, wind direction and speed, relative humidity, and **barometric pressure.**

OBJECTIVE: Learn the fun and value of keeping weather data.

Weather Tote:

Weather instruments that you made (see activity, Making Weather Instruments) or purchased. Choose the weather instruments you wish to use.

- Rain gauge
- Wind vane
- Anemometer
- Barometer
- Hygrometer
- Thermometer

Do it

See the Making Weather Instruments activity in this manual for instructions to make many of the instruments listed in the Weather Tote.

1 Set up your weather instruments where they will not be disturbed and away from trees and buildings that might affect wind direction and speed. Your thermometer, hygrometer, and barometer must be in the shade.

2 Make a Weather Data Sheet with enough rows for one month. (See example in Table 15.) Include a column for the date and each data type that you wish to collect: precipitation, temperature, wind (direction and speed), barometric pressure, and **relative humidity.**

3 Title your Weather Data Sheet with the month and year that you will be recording data.

4 Record weather data for a month using your weather instruments.

SHARE WHAT HAPPENED: Was your weather station in a good place? Why, or why not?

APPLY: How might you use your weather data?

GENERALIZE TO YOUR LIFE: What do you think you could learn by averaging weather data from 30 years of records?

Fly Higher

Learn about CoCoRaHS—the **Co**mmunity **Co**llaborative **Ra**in, **H**ail & **S**now Network. A brief overview is given below. Learn more by visiting www.cocorahs.org or reading the *Be a Weather Monitor* activity in the Level 3.

- CoCoRaHS is a network of **grassroots** volunteers who are dedicated to measuring and recording the rain, hail, and snow from their backyards.
- CoCoRaHS is the largest provider of daily precipitation observations in the country.
- Organizations involved in agriculture, public safety, and natural-resources management find great benefit from the data the network volunteers collect.
- Data is available free of charge at the Network's website, www.cocorahs.org.

TABLE 15. **WEATHER DATA SHEET FOR MONTH, YEAR**

Date	Temperature	Precipitation	Wind Speed/ Wind Direction	Cloud Conditions	Air Pressure	Humidity	Other

GLOSSARY

Absorb: To take in all or part of a substance: The earth's surfaces absorb radiant energy.

Aerosol: Particles suspended in the air.

Air mass: A large body of air with similar temperature and moisture characteristics.

Air pressure: The force that air exerts on any surface in contact with it.

Albedo: The ratio of reflected light energy to absorbed light energy; ability to reflect sunlight.

Albedo effect: A measure of how much solar radiation is reflected.

Atmosphere: A layer of gases surrounding the earth, held in place by gravity.

Barometric pressure: Atmospheric pressure as measured by a barometer, usually in millibars or inches of mercury.

Calibrate: To check and adjust a measurement precisely against a standard measure.

Climograph: A graphic depiction of the precipitation and temperature for a certain location.

Clockwise: The rotational direction of the hands of a clock.

Cold front: A front formed by a cold air mass pushing into an area of warmer air.

Concentration: The amount of a substance in a given amount of another substance.

Condense: To convert a vapor or gas to a liquid.

Conduction: Energy transferred directly from one object to another through contact.

Contract: To become drawn together or reduced; shrink.

Contrail: Water droplets or ice crystals condensing in the supersaturated air of the upper atmosphere as a result of an aircraft's passage.

Convection: The transfer of energy through circulation within a liquid or gas.

Counterclockwise: Opposite to the rotational direction of the hands of a clock.

Deflect: To turn aside from a course.

Density: The degree of compactness of a substance; mass/unit measure ~ weight.

Dew-point temperature: The temperature to which air must be cooled to reach the point of saturation of water vapor; the temperature at which dew forms.

El Niño: A band of warm ocean water temperatures that periodically develops off the western coast of South America. It can cause climatic changes such as floods and droughts in many regions of the world, particularly along the coasts of Chile, Peru, New Zealand, and Australia.

Ellipse: A closed, symmetric curve shaped like an oval.

Equilibrium: A state of balance between opposing forces or actions.

Equinox: The time when the sun crosses over the equator, making night and day approximately equal in length all over the earth.

Evaporate: To convert a liquid to a gas or vapor.

Evapotranspiration: The combination of water being evaporated and transpired.

Fluid: A substance, gas or liquid, that is capable of flowing.

Fossil fuel: Any naturally occurring carbon or hydrocarbon fuel, such as coal, petroleum, peat, and natural gas, formed by the decomposition of prehistoric organisms.

Front: A boundary at ground level between two different air masses as one air mass moves into the area of the other.

Grassroots: Used to describe organizations that begin and operate at the local or community level.

Greenhouse gases: Gases in the earth's atmosphere that absorb and emit radiation, or heat energy, in the thermal infrared range.

Hardiness: The ability of a plant to withstand cold temperatures.

Heat index: An addition to temperature data during hot and humid weather to indicate health risks more accurately than temperature alone.

Infrared radiation: A form of longwave radiation, or heat energy, given off by warm objects.

Isobar: A line on a weather map connecting points of equal barometric pressure.

Isoline: A line on a map or chart connecting points of equal value.

Isotherm: A line on a weather map connecting points of equal temperature.

Jet stream: Fast-flowing, narrow currents of air. The main jet streams are located near the tropopause.

La Niña: A cold phase in which the sea surface temperature across the equatorial eastern central Atlantic Ocean is lower than normal by 3–5°C. La Niña often follows El Niño.

Latent heat: Heat absorbed or radiated during a phase change.

Longwave radiation: Radiant energy emitted by lower energy objects, like a warm body giving off heat.

Mammatus: An ominous-looking cloud forming at the bottom of cumulonimbus clouds, with a sagging, pouch-like appearance.

Overwinter: To survive freezing conditions, often applied to perennial plants.

Precipitation: The amount of water that falls to earth as hail, drizzle, rain, sleet, or snow.

Radiant energy: Energy emitted in waves.

Relative humidity: The ratio, given as a percentage, of the amount of water vapor in the air compared to the amount of water vapor in saturated air.

Saturation: An atmospheric condition at which the moisture level is at 100% relative humidity.

Scale: A proportion that shows the relative size between objects but that is not the same size as the original objects.

Shortwave radiation: Radiant energy emitted by high-energy sources like the sun or a light bulb, which emit light energy.

Solar maximum: A time of great solar activity identified by the occurrence of many sunspots.

Solar minimum: A time of low solar activity identified by the occurrence of few sunspots.

Solstice: Either of two times a year when the sun is at its greatest distance north or south of the equator.

Sunspot: Dark, cool patches that appear on the surface of the sun.

Sunspot cycle: An approximately 11-year cycle of the occurrence of sunspots.

Thermal: A rising warm current of air caused by uneven heating of the surface.

Tropopause: The transition between the troposphere (where temperature decreases with altitude) and the stratosphere (where temperature increases with altitude).

Troposphere: The lowest portion of the earth's atmosphere.

Vortex: A whirling mass of water or air that forms a visible column or spiral.

Water vapor: Water in the form of an invisible, colorless gas.

Weather map: A daily report of observations and probable changes in the weather.

Weather station: One of 1,500 locations where the National Oceanic and Atmospheric Administration (NOAA) collects weather data.

Weather station model: A symbolic illustration showing weather conditions at a weather station recorded using abbreviations and symbols.

Windchill: Addition to temperature data during cold and windy weather to indicate health risks more accurately than temperature alone.

Zone: An area where the average annual minimum temperature is within a set temperature range.

APPENDIX

RELATIVE HUMIDITY AND DEWPOINT CHART

Top Number–Relative Humidity in %; FF0000Bottom Number–Dew Point Temperature in ˚F000000)01+000

Temperature Depression (Difference between Dry and Wet Bulb Thermometer Readings, Degrees Fahrenheit)

Each cell shows: Relative Humidity / Dew Point

Dry Bulb F	1	2	3	4	5	6	7	8	9	10	11	12	13	14	15	16	17	18	19	20	21	22	23	24	25	26
40	92 / 37	83 / 34	75 / 31	66 / 28	60 / 26	52 / 23	45 / 20	37 / 17	29 / 14	22 / 12	15 / 9	7 / 7														
41	92 / 38	84 / 35	76 / 32	69 / 30	61 / 27	54 / 24	46 / 22	39 / 19	31 / 16	24 / 14	17 / 11	10 / 9	3 / 8													
42	92 / 39	85 / 37	77 / 34	69 / 31	62 / 28	55 / 26	47 / 23	40 / 20	33 / 18	26 / 15	19 / 13	12 / 10	5 / 8													
43	92 / 40	85 / 38	77 / 35	70 / 32	63 / 30	55 / 27	48 / 24	42 / 22	35 / 20	28 / 17	21 / 15	14 / 12	8 / 10	1 / 7												
44	93 / 41	85 / 39	78 / 36	71 / 34	63 / 31	56 / 28	49 / 26	43 / 23	36 / 21	30 / 19	23 / 16	16 / 14	10 / 12	4 / 9												
45	93 / 42	86 / 40	78 / 37	71 / 35	64 / 32	57 / 30	51 / 27	44 / 25	38 / 23	31 / 20	25 / 18	18 / 15	12 / 13	6 / 11												
46	93 / 43	86 / 41	79 / 38	72 / 36	65 / 33	58 / 31	52 / 29	45 / 26	39 / 24	32 / 22	26 / 19	20 / 17	14 / 15	8 / 13	2 / 11											
47	93 / 44	86 / 42	79 / 39	72 / 37	66 / 35	59 / 32	53 / 30	46 / 28	40 / 25	34 / 23	28 / 21	22 / 19	16 / 17	10 / 15	5 / 13											
48	93 / 45	86 / 43	79 / 40	73 / 38	66 / 36	60 / 34	54 / 31	47 / 29	41 / 27	35 / 25	29 / 22	23 / 20	18 / 18	12 / 16	7 / 14	1 / 12										
49	93 / 46	86 / 44	80 / 42	73 / 39	67 / 37	61 / 35	54 / 32	48 / 30	42 / 28	36 / 26	31 / 24	25 / 22	19 / 20	14 / 18	9 / 16	3 / 14										
50	93 / 47	87 / 45	80 / 43	74 / 41	67 / 38	61 / 36	55 / 34	49 / 32	43 / 29	38 / 28	32 / 26	27 / 24	21 / 22	16 / 20	10 / 18	5 / 16										
51	94 / 49	87 / 46	81 / 44	75 / 42	68 / 39	62 / 37	56 / 35	50 / 33	45 / 31	39 / 29	34 / 27	28 / 25	23 / 23	17 / 21	12 / 19	7 / 18	2 / 16									
52	94 / 50	87 / 47	81 / 45	75 / 43	69 / 41	63 / 39	57 / 37	51 / 34	46 / 33	40 / 30	35 / 29	29 / 26	24 / 25	19 / 23	14 / 21	9 / 19	4 / 17									
53	94 / 51	87 / 48	81 / 46	75 / 44	69 / 42	63 / 40	58 / 38	52 / 36	47 / 34	41 / 32	36 / 30	31 / 28	26 / 26	20 / 24	16 / 23	10 / 21	6 / 19	1 / 17								
54	94 / 52	88 / 50	82 / 48	76 / 45	70 / 43	64 / 41	59 / 39	53 / 37	48 / 35	42 / 33	37 / 31	32 / 30	27 / 28	22 / 26	17 / 24	12 / 22	8 / 21	3 / 19								
55	94 / 53	88 / 51	82 / 49	76 / 46	70 / 44	65 / 42	59 / 40	54 / 38	49 / 37	43 / 34	38 / 33	33 / 31	28 / 29	23 / 27	19 / 26	14 / 24	9 / 22	5 / 21								
56	94 / 54	88 / 52	82 / 50	76 / 47	71 / 46	65 / 43	60 / 42	55 / 40	50 / 38	44 / 36	39 / 34	34 / 32	30 / 31	25 / 29	20 / 27	16 / 26	11 / 24	7 / 23	2 / 21							
57	94 / 55	88 / 53	82 / 51	77 / 49	71 / 47	66 / 45	61 / 43	55 / 41	50 / 39	45 / 37	40 / 35	35 / 34	31 / 32	26 / 30	22 / 29	17 / 27	13 / 26	8 / 24	4 / 22							
58	94 / 56	88 / 54	83 / 52	77 / 50	72 / 48	66 / 46	61 / 44	56 / 42	51 / 40	46 / 39	41 / 37	37 / 35	32 / 34	27 / 32	23 / 30	18 / 28	14 / 27	10 / 26	6 / 24	1 / 22						
59	94 / 57	89 / 55	83 / 53	78 / 51	72 / 49	67 / 47	62 / 45	57 / 44	52 / 42	47 / 40	42 / 38	38 / 37	33 / 35	29 / 33	24 / 32	20 / 30	16 / 29	11 / 27	7 / 26	3 / 24						
60	94 / 58	89 / 56	83 / 54	78 / 52	73 / 50	68 / 48	63 / 47	58 / 45	53 / 43	48 / 41	43 / 39	39 / 38	34 / 36	30 / 35	26 / 33	21 / 32	17 / 30	13 / 29	9 / 27	5 / 26	1 / 24					
61	94 / 59	89 / 57	84 / 55	78 / 53	73 / 51	68 / 49	63 / 48	58 / 46	54 / 44	49 / 43	44 / 41	40 / 39	35 / 38	31 / 36	27 / 35	22 / 33	18 / 31	14 / 30	10 / 29	7 / 28	3 / 26					
62	94 / 60	89 / 58	84 / 56	79 / 54	74 / 53	69 / 51	64 / 49	59 / 47	54 / 45	50 / 44	45 / 42	41 / 41	36 / 39	32 / 38	28 / 36	24 / 35	20 / 33	16 / 32	12 / 30	8 / 29	4 / 27	1 / 26				
63	95 / 61	89 / 59	84 / 57	79 / 55	74 / 54	69 / 52	64 / 50	60 / 49	55 / 47	50 / 45	46 / 44	42 / 42	37 / 40	33 / 39	29 / 37	25 / 36	21 / 35	17 / 33	13 / 32	10 / 31	6 / 29	2 / 28				
64	95 / 62	90 / 60	84 / 58	79 / 56	74 / 55	70 / 53	65 / 51	60 / 50	56 / 48	51 / 46	47 / 45	43 / 43	38 / 42	34 / 40	30 / 39	26 / 37	22 / 36	18 / 34	15 / 33	11 / 32	7 / 31	4 / 29				
65	95 / 63	90 / 61	85 / 60	80 / 58	75 / 56	70 / 54	66 / 53	61 / 51	56 / 49	52 / 48	48 / 46	44 / 45	39 / 43	35 / 42	31 / 40	27 / 39	24 / 38	20 / 36	16 / 35	12 / 33	9 / 32	5 / 31	2 / 30			
66	95 / 64	90 / 62	85 / 61	80 / 59	75 / 57	71 / 56	66 / 54	61 / 52	57 / 51	53 / 49	48 / 47	44 / 46	40 / 44	36 / 43	32 / 42	29 / 40	25 / 39	21 / 38	17 / 36	14 / 35	10 / 34	7 / 33	3 / 31			
67	95 / 65	90 / 63	85 / 62	80 / 60	75 / 58	71 / 57	66 / 55	62 / 53	58 / 52	53 / 50	49 / 49	45 / 47	41 / 40	37 / 44	33 / 43	30 / 42	26 / 40	22 / 30	19 / 38	15 / 36	12 / 35	8 / 34	5 / 33	2 / 32		
68	95 / 66	90 / 64	85 / 63	80 / 61	76 / 59	71 / 58	67 / 56	62 / 54	58 / 53	54 / 51	50 / 50	46 / 49	42 / 47	38 / 46	34 / 44	31 / 43	27 / 42	23 / 40	20 / 39	16 / 38	13 / 37	10 / 36	6 / 34	3 / 33		
69	95 / 67	90 / 65	85 / 64	81 / 62	76 / 60	72 / 59	67 / 57	63 / 56	59 / 54	55 / 53	51 / 51	47 / 50	43 / 48	39 / 47	35 / 46	32 / 45	28 / 43	24 / 42	21 / 41	18 / 39	14 / 38	11 / 37	8 / 36	5 / 35	1 / 33	
70	95 / 68	90 / 66	86 / 65	81 / 63	77 / 62	72 / 60	68 / 58	64 / 57	59 / 55	55 / 54	51 / 52	48 / 51	44 / 50	40 / 48	36 / 47	33 / 46	29 / 44	25 / 43	22 / 42	19 / 41	15 / 39	12 / 38	9 / 37	6 / 36	3 / 35	
71	95 / 69	90 / 67	86 / 66	81 / 64	77 / 63	72 / 61	68 / 59	64 / 58	60 / 57	56 / 55	52 / 54	48 / 52	45 / 51	41 / 50	37 / 48	33 / 47	30 / 46	27 / 45	23 / 43	20 / 42	17 / 41	13 / 40	10 / 39	7 / 38	4 / 36	1 / 35
72	95 / 70	91 / 69	86 / 67	82 / 66	77 / 64	73 / 62	69 / 61	65 / 59	61 / 58	57 / 57	53 / 55	49 / 54	45 / 52	42 / 51	38 / 50	34 / 48	31 / 47	28 / 46	24 / 45	21 / 44	18 / 42	15 / 41	12 / 40	9 / 39	6 / 38	3 / 37
73	95 / 71	91 / 70	86 / 68	82 / 67	78 / 65	73 / 63	69 / 62	65 / 60	61 / 59	57 / 58	53 / 56	50 / 55	46 / 54	42 / 52	39 / 51	35 / 50	32 / 49	29 / 47	25 / 46	22 / 45	19 / 44	16 / 43	13 / 42	10 / 41	7 / 40	4 / 38
74	95 / 72	91 / 71	86 / 69	82 / 68	78 / 66	74 / 65	69 / 63	65 / 61	61 / 60	58 / 59	54 / 57	50 / 56	47 / 55	43 / 53	39 / 52	36 / 51	33 / 50	29 / 48	26 / 47	23 / 46	20 / 45	17 / 44	14 / 43	11 / 42	8 / 41	5 / 40
75	96 / 74	91 / 72	86 / 70	82 / 69	78 / 67	74 / 66	70 / 64	66 / 63	62 / 61	58 / 60	54 / 58	51 / 57	47 / 56	44 / 55	40 / 53	37 / 52	34 / 51	30 / 50	27 / 49	24 / 48	21 / 47	18 / 45	15 / 44	12 / 43	9 / 42	7 / 42
76	96 / 75	91 / 73	87 / 71	82 / 70	78 / 68	74 / 67	70 / 65	66 / 64	62 / 62	59 / 61	55 / 60	51 / 58	48 / 57	44 / 56	41 / 55	38 / 54	34 / 52	31 / 51	28 / 50	25 / 49	22 / 48	19 / 47	16 / 46	13 / 45	11 / 44	8 / 43
77	96 / 76	91 / 74	87 / 72	83 / 71	79 / 69	74 / 68	71 / 67	67 / 65	63 / 64	59 / 62	56 / 61	52 / 60	48 / 58	45 / 57	42 / 56	39 / 55	35 / 54	32 / 53	29 / 51	26 / 50	23 / 49	20 / 48	17 / 47	14 / 46	12 / 45	9 / 44
78	96 / 77	91 / 75	87 / 73	83 / 72	79 / 70	75 / 69	71 / 68	67 / 66	63 / 65	60 / 64	56 / 62	53 / 61	49 / 60	46 / 59	43 / 57	39 / 56	36 / 55	33 / 54	30 / 53	27 / 52	24 / 51	21 / 50	18 / 48	16 / 48	13 / 47	10 / 46
79	96 / 78	91 / 76	87 / 74	83 / 73	79 / 71	75 / 70	71 / 69	68 / 67	64 / 66	60 / 65	57 / 64	53 / 62	50 / 61	46 / 60	43 / 58	40 / 57	37 / 56	34 / 55	31 / 54	28 / 53	25 / 52	22 / 51	19 / 50	17 / 49	14 / 48	11 / 47
80	96 / 79	91 / 77	87 / 75	83 / 74	79 / 72	75 / 71	72 / 70	68 / 68	64 / 67	61 / 66	57 / 65	54 / 63	50 / 62	47 / 61	44 / 60	41 / 59	38 / 57	35 / 57	32 / 56	29 / 54	26 / 53	23 / 52	20 / 51	18 / 50	15 / 49	12 / 48
82	96 / 81	92 / 79	88 / 78	84 / 76	80 / 75	76 / 73	72 / 72	69 / 71	65 / 69	61 / 68	58 / 67	55 / 66	51 / 64	48 / 63	45 / 62	42 / 61	39 / 60	36 / 59	33 / 58	30 / 57	28 / 56	25 / 55	22 / 54	20 / 53	17 / 52	14 / 51
84	96 / 83	92 / 81	88 / 80	84 / 78	80 / 77	76 / 75	73 / 74	69 / 73	66 / 72	62 / 70	59 / 69	56 / 68	52 / 67	49 / 66	46 / 65	43 / 63	40 / 62	37 / 61	35 / 61	32 / 60	29 / 58	26 / 57	24 / 57	21 / 56	19 / 55	16 / 54
86	96 / 85	92 / 83	88 / 82	84 / 80	81 / 79	77 / 78	73 / 76	70 / 75	66 / 74	63 / 73	60 / 72	57 / 71	53 / 69	50 / 68	47 / 67	44 / 66	42 / 65	39 / 64	36 / 63	33 / 62	31 / 61	28 / 60	26 / 59	23 / 58	21 / 58	18 / 56
88	96 / 87	92 / 85	88 / 84	85 / 83	81 / 81	77 / 80	74 / 79	70 / 77	67 / 76	64 / 75	61 / 74	57 / 73	54 / 71	51 / 70	48 / 69	46 / 69	43 / 67	40 / 66	37 / 65	35 / 65	32 / 64	30 / 63	27 / 62	25 / 61	22 / 60	20 / 59
90	96 / 89	92 / 87	89 / 86	85 / 85	81 / 83	78 / 82	74 / 81	71 / 80	68 / 78	65 / 77	61 / 76	58 / 75	55 / 74	52 / 73	49 / 72	47 / 71	44 / 70	41 / 69	39 / 68	36 / 67	34 / 66	31 / 65	29 / 64	26 / 63	24 / 63	22 / 62
92	96 / 91	92 / 89	89 / 88	85 / 87	82 / 86	78 / 84	75 / 83	72 / 82	68 / 80	65 / 79	62 / 78	59 / 77	56 / 76	53 / 75	50 / 74	48 / 73	45 / 72	42 / 71	40 / 70	37 / 69	35 / 69	32 / 68	30 / 67	28 / 66	25 / 65	23 / 64
94	96 / 93	93 / 91	89 / 90	85 / 89	82 / 88	79 / 86	75 / 85	72 / 84	69 / 83	66 / 83	63 / 82	60 / 81	57 / 80	54 / 79	51 / 77	49 / 76	46 / 76	43 / 75	41 / 73	38 / 73	36 / 72	33 / 71	31 / 70	29 / 69	27 / 68	24 / 67
96	96 / 95	93 / 93	89 / 92	86 / 91	82 / 90	79 / 88	76 / 87	73 / 86	69 / 85	66 / 84	63 / 83	61 / 82	58 / 81	55 / 80	52 / 79	50 / 78	47 / 77	44 / 76	42 / 75	39 / 74	37 / 73	35 / 73	32 / 72	30 / 71	28 / 70	26 / 69

Atlantic Basin Hurricane Tracking Chart

National Hurricane Center, Miami, Florida

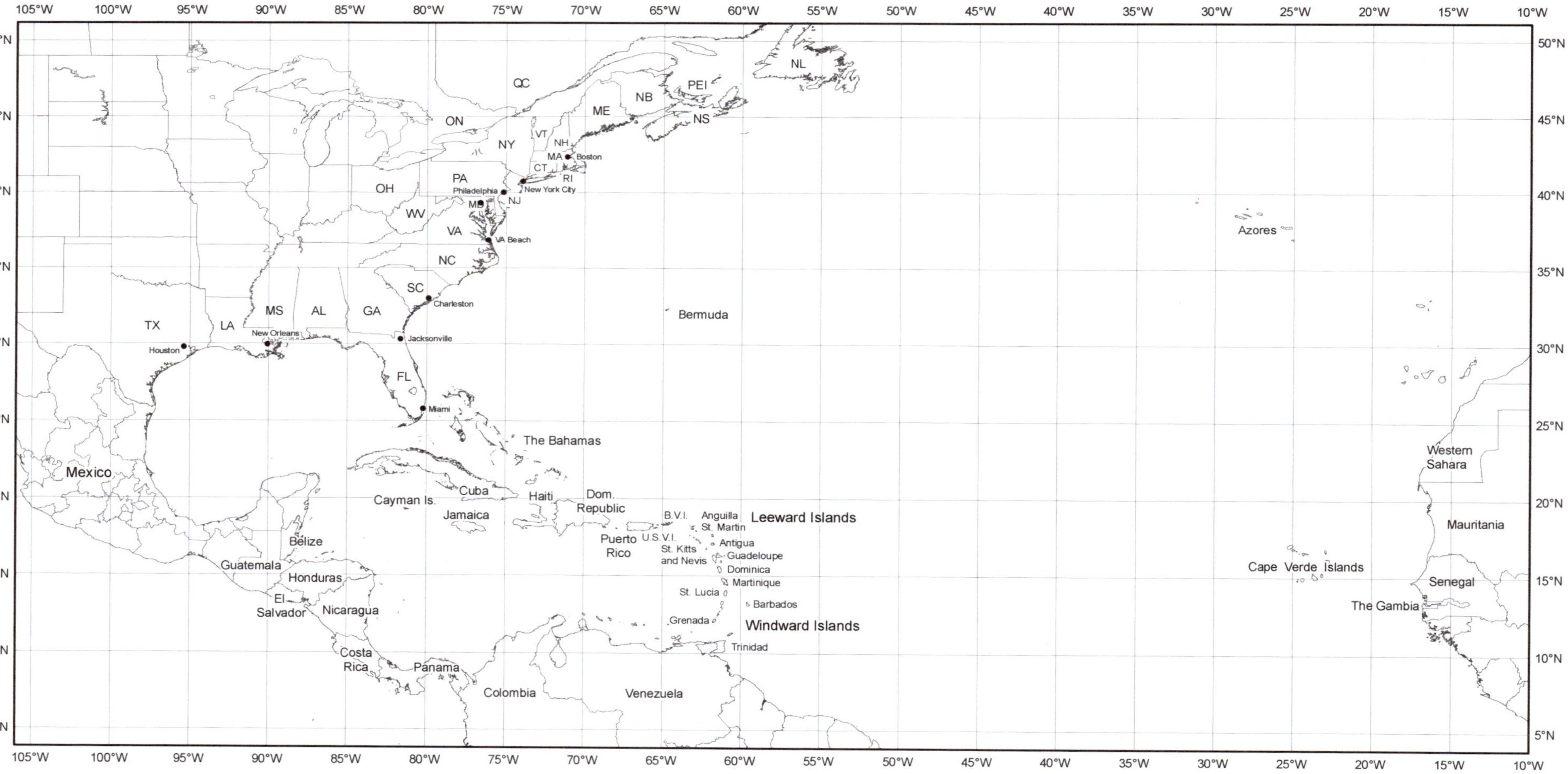

Build Your Own Weather Station

"I believe there is a train under here somewhere!"

Standing tall on North Dakota snow. A March blizzard nearly buried utility poles. Caption jokingly read "I believe there is a train under here somewhere!" Courtesy Dr. Herbert Kroehl, NGDC

Every year, thousands of lives and millions of dollars are saved by severe weather warnings from the National Weather Service. From its earliest beginnings (on February 9th, 1870), the primary mission of the National Weather Service has been to protect life and property by providing information about dangerous weather conditions. Originally, the National Weather Service was called "The Division of Telegrams and Reports for the Benefit of Commerce" and was part of the U.S. Army. Later, its name was shortened to the Weather Bureau and it became part of the Department of Agriculture, then the Department of Commerce.

The first "weathermen" were "observing-sergeants" of the Army's Signal Service Corps. Weather forecasting in those early years was based almost entirely on the assumption that the weather observed in one location on a particular day would move to downwind locations on following days. Today, satellites, computers, and a variety of scientific instruments are added to this basic assumption to make accurate weather predictions and provide warnings about dangerous weather.

Here's how you can make your own weather observation station!

What You Will Do

Build six instruments that you can use to make scientific measurements of your local weather

Build an Anemometer to Measure Wind Speed

What You Will Need

- ❒ Five paper cups - Three ounce size
- ❒ Two straight plastic soda straws
- ❒ Straight pin
- ❒ Paper punch
- ❒ Stapler
- ❒ Sharp pencil with eraser
- ❒ Felt tip marker
- ❒ Watch or timer

Warning

Be careful with the straight pin!

How to Do It:

1. Using a paper punch, punch a hole in four paper cups about 1/2-inch below the rim of the cups.

2. Punch four equally spaced holes in a fifth paper cup about 1/4-inch below the rim, and a fifth hole in the center of the bottom of the cup (you will probably need to use the pencil to make the hole in the bottom).

3. Push a soda straw through the hole in one of the first four cups. Flatten the end of the straw and staple it to the side of the cup opposite the hole. Repeat this step with the other straw and another of the first four cups.

4. Slide one of the cup and straw assemblies through two opposite holes in the side of the fifth cup. Push another one-hole cup onto the straw, and turn this cup so that the open ends of the two cups on the straw face in opposite directions. Flatten the end of the straw, and staple it to the side of the second cup. Measure the distance between the centers of the two cups. This is the diameter of your anemometer.

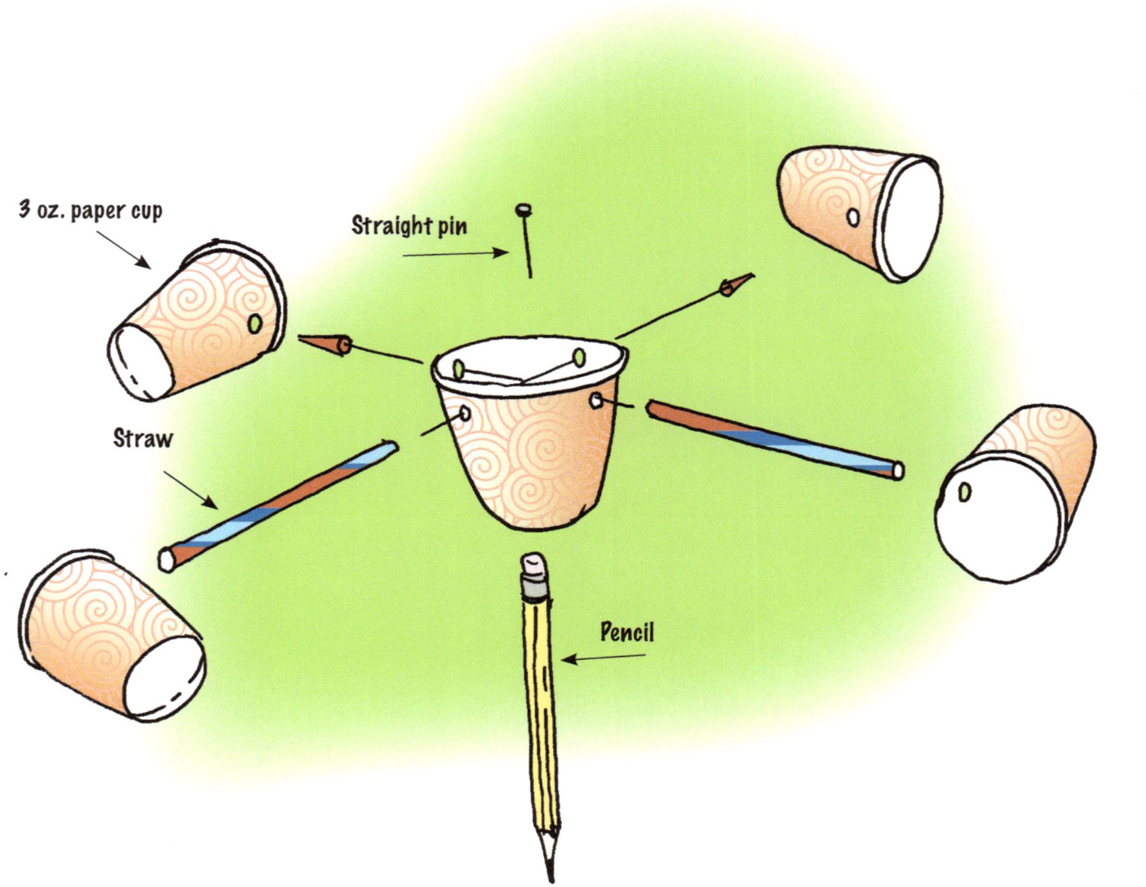

National Oceanic and Atmospheric Administration (NOAA)

5. Repeat Step 4 with the remaining cup and straw assembly and the remaining one-hole cup. Before stapling the end of the straw to the last cup, turn the cups so that the open end of each cup faces the closed end of the next cup.

6. Adjust the cup and straw assemblies so that they are centered inside the fifth cup. Push the straight pin through the two straws where they intersect.

7. Push the eraser end of the pencil through the hole in the bottom of the fifth cup, and push the straight pin into the eraser as far as it will go. Now your anemometer is ready to use.

8. To use the anemometer, hold the pencil vertically in a wind, and count the number of revolutions per minute (use the felt tip marker to make a mark on one cup so that you can easily see when the cup has travelled through one complete revolution). To convert revolutions per minute (rpm) into approximate wind speed:

   a. Multiply rpm by the diameter (in inches) of your anemometer (measured in Step 4)
   b. Multiply the result by 0.003. This is the approximate wind speed in miles per hour.

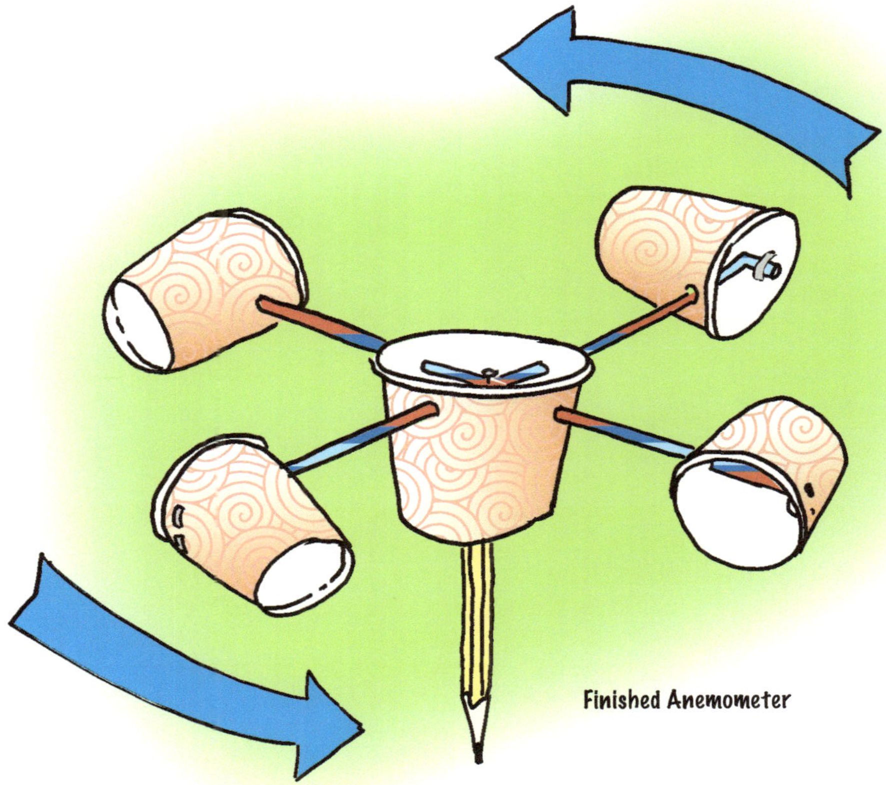

Finished Anemometer

This calculation does not give exact wind speed, because drag, friction, and other forces also affect the speed at which your anemometer rotates.

Build a Weather Vane to Find Wind Direction

What You Will Need

- ❒ Broomstick or long wooden dowel, about one inch diameter
- ❒ Aluminum baking dish, about six inches x nine inches
- ❒ Wood stick, about 3/4 inch square and 12 inches long
- ❒ Nail, about one inch long
- ❒ Metal washer with a hole slightly larger than the nail
- ❒ Duct tape
- ❒ Small saw or serrated knife
- ❒ Scissors strong enough to cut the aluminum baking dish
- ❒ Ruler or tape measure
- ❒ Silicone or other glue that will stick to aluminum
- ❒ Leather gloves
- ❒ (Optional) Hand drill, and small drill bit slightly larger than the nail

Warning

Be careful of the sharp edges on the pieces of cut aluminum! Use gloves to protect your hands until the edges are taped.

How to Do It

1. Use the saw or serrated knife to cut a notch about 1/2-inch deep into each end of the wood stick. The notches shuld be parallel (see drawing on page 00).

2. Rotate the stick so that the two slots are vertical. Use the ruler or tape measure to find the exact center of the wood stick.

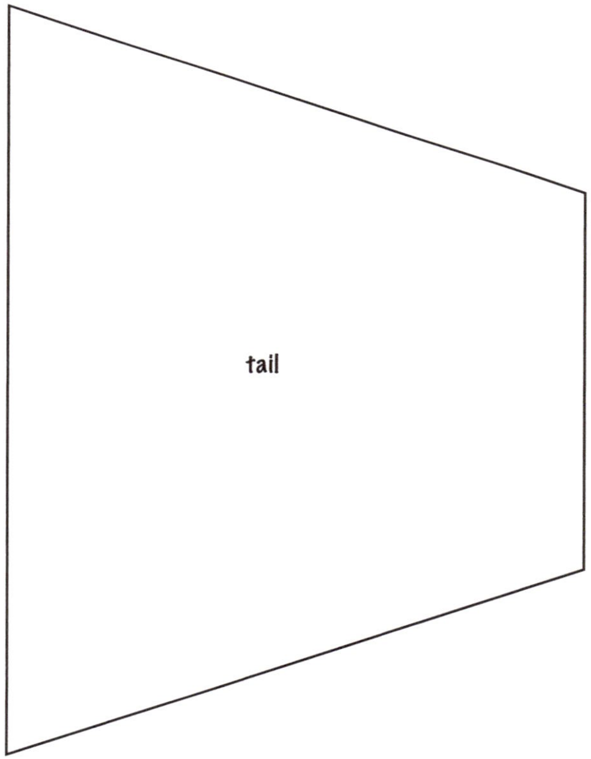

Mark this spot on the upper surface of the stick, and drive a nail through the marked spot. Be careful: if the nail is too big, the stick will probably split. To avoid this, drill a hole slightly larger than the nail through the marked spot. You may need an adult to help with the drilling.

3. Cut the head and tail pieces of the Weather Vane from the aluminum baking dish using the pattern as a guide. **Be Careful—The Edges Are Sharp!** Use duct tape to cover the sharp edges.

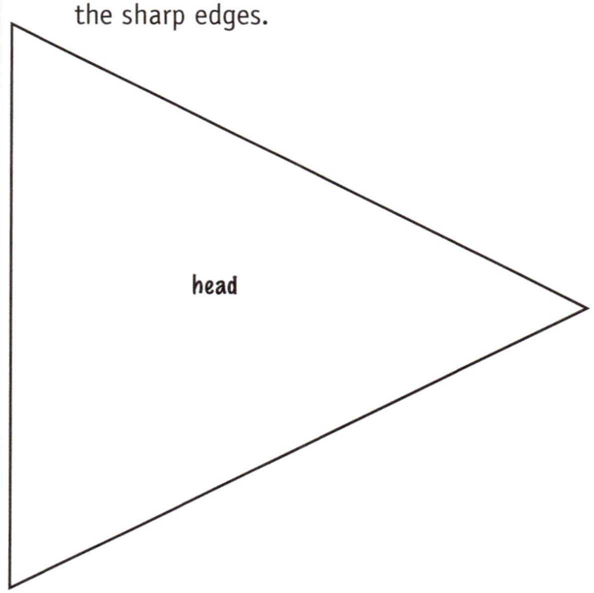

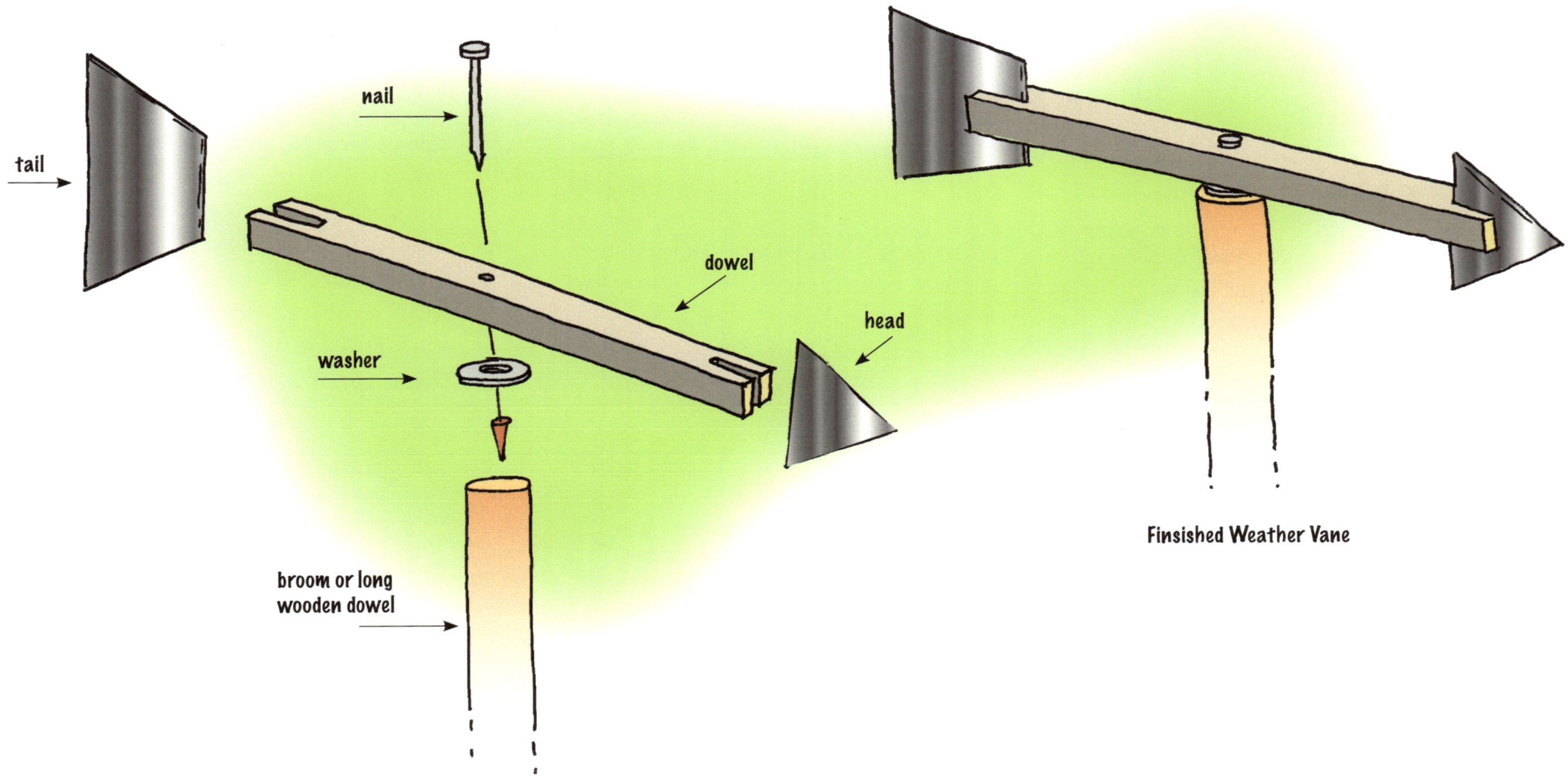

4. Fit the head piece into one of the slots in the wood stick and fit the tail piece into the other slot. Glue the head and tail pieces into place and allow the glue to dry.

5. Attach the Weather Vane to the broomstick or dowel, by placing the washer on one end of the dowel and hammering the nail through the wooden stick into the dowel. Be sure the stick still moves freely around the nail.

6. Mount your Weather Vane outside where there are no nearby obstructions to block the wind. Try to get the dowel as high as you can while still keeping it steady and secure.

Winds are named according to the direction from which the wind is blowing, so a "north wind" is blowing from the north. The head of the Weather Vane will point to the direction from which the wind is blowing.

National Oceanic and Atmospheric Administration (NOAA)

Build a Barometer for Measuring Atmospheric Pressure

What You Will Need

- ❒ 12-inch ruler
- ❒ Drinking glass or other container with sides tall enough to support the ruler
- ❒ Clear plastic drinking straw or piece of clear plastic tubing, about 12-inches long
- ❒ Modeling clay or chewing gum
- ❒ Clear tape
- ❒ (Optional) Food coloring

How to Do It

1. Tape the plastic straw or plastic tubing to the ruler so that one end is lined up with the "1/2-inch" mark on the ruler.

2. Stand the ruler-tubing assembly upright in the glass (or other container), and tape the assembly to the top of the container.

3. Fill the container about 3/4-full of water. If you want colored water, first mix food coloring with the water in another container.

4. Use the modeling clay or chewing gum (you'll have to chew it until it is soft enough) to plug the end of the straw or plastic tubing near the top of the ruler.

5. Carefully pour out some of the water so the container is about half full. Be sure the lower end of the straw or tubing stays beneath the water surface while you do this! When you are finished, the water in the straw or tube should be higher than the water in the container. Your barometer is now finished. Since barometers are sensitive to minor changes in weather conditions, keep your barometer indoors for greatest accuracy.

6. Keep a daily record of the height of the water in the tube, using the scale on the ruler. The water level in the tube will rise and fall as atmospheric pressure changes. When atmospheric pressure increases, air presses on the surface of the water in the container causing the height of the water in the tube to rise. When atmospheric pressure decreases, there is less pressure on the surface of the water in the container so the height of the water in the tube falls. Decreasing atmospheric pressure usually indicates that a low pressure area is approaching, and this often brings clouds and rain. Increasing atmospheric pressure often indicates fair weather.

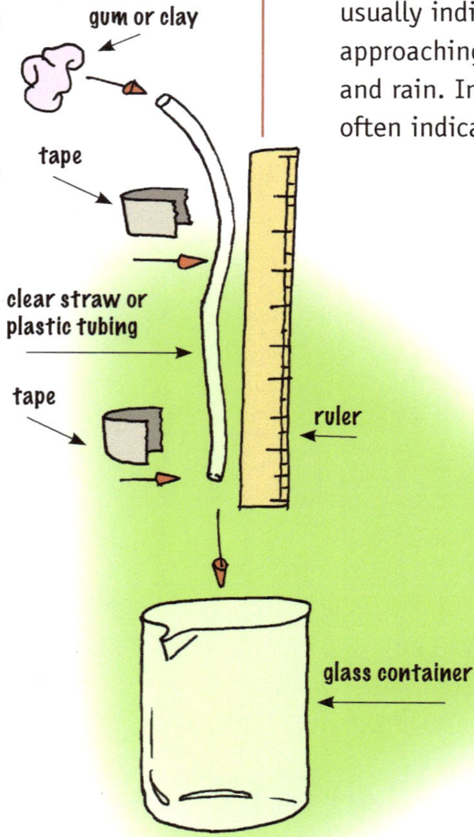

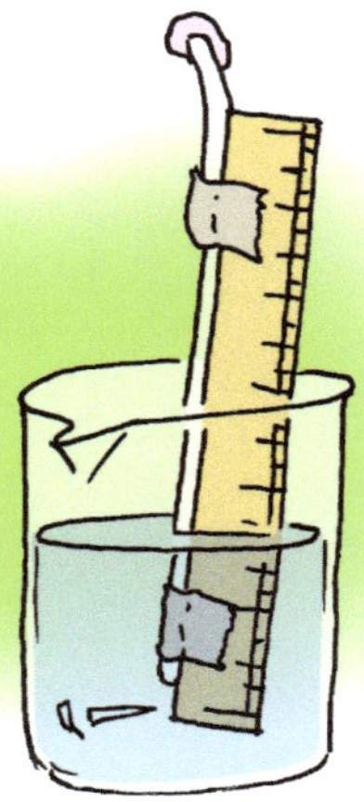

Finished Barometer

National Oceanic and Atmospheric Administration (NOAA)

Build a Screened Thermometer to Measure Air Temperature

What You Will Need

- ❒ A wooden or plastic box, large enough to hold the thermometer and your hygrometer; see Step 1 under "How to Do It"
- ❒ Thermometer, about 0°F to 120°F
- ❒ White paint and paint brush
- ❒ Nails, screws, glue, or tape to attach the thermometer to the box

How to Do It

1. The wood or plastic box is supposed to protect your weather instruments from wind, rain, and direct sun, but still allow air to circulate so the instruments can get accurate readings. A box with a hinged lid that can be turned on its side is perfect. Turn the box on its side, and cut several slots near what is now the bottom of the box. Paint the outside of the box with white paint, and find a safe, shady outdoor location. The north side of buildings has the most shade. Try to find a location that is three to four feet above the ground.

2. Attach the thermometer to the back of the box with tape, glue, screws, or nails. The bulb of the thermometer should be about two inches above the bottom of the box.

Build a Hygrometer to Measure Humidity

What You Will Need

- ❒ Piece of wood or styrofoam about nine inches long and four inches wide
- ❒ Flat piece of plastic, thin enough to cut with scissors; about three inches long and one inch wide (an old credit card or laminated luggage tag works well)
- ❒ Two small nails
- ❒ Three strands of human hair, about eight inches long
- ❒ Dime
- ❒ Glue
- ❒ Tape
- ❒ Hammer
- ❒ Scissors

How to Do It

1. Cut the plastic into a pointer as shown on the pattern below.

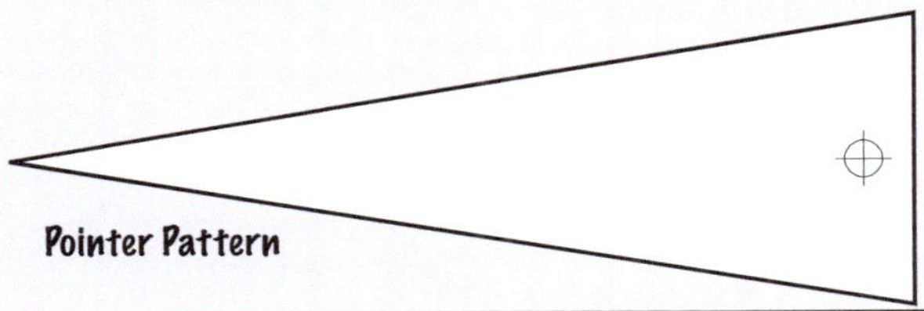

Pointer Pattern

2. Poke one of the nails through the pointer near the base of the triangle. Wiggle the nail around until the pointer moves freely and loosely around the nail.

3. Tape the dime onto the pointer near the tip of the triangle.

4. Glue the hair strands onto plastic between the nail hole and the dime.

5. Use a nail to fasten the pointer to the wood or styrofoam base about 3/4 of the way down the side. Be sure the pointer can still turn freely on the nail.

6. Attach the other nail to the base about one inch from the top of the base, in line with the spot where the hair is glued to the pointer.

7. Pull the free ends of the hair tight so that the pointer is horizontal. Wrap the hair

around the upper nail and glue to hold the hair in place.

8. Make a photocopy of the scale and cut it out. Glue the scale to the base so that the pointer is pointed to the "0" mark. Your hygrometer is finished!

9. Human hair will expand and lengthen when the air is moist, causing the pointer to move down. When the air is dry, the hair will contract and shorten, causing the pointer to move up. Use the scale to record the pointer's position. Keep your hygrometer in a sheltered location. The box used for the screened thermometer is ideal.

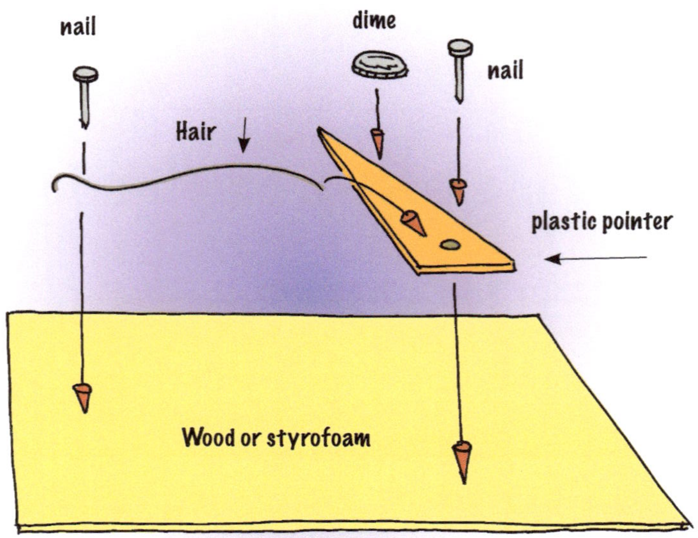

Finished Hygrometer

National Oceanic and Atmospheric Administration (NOAA)

Build a Rain Gauge to Measure Rainfall

What You Need

- ❐ Straight-sided glass or plastic container, with a diameter of about two inches or less (such as an olive jar)
- ❐ Coat hanger or wire bent to make a holding rack (see picture)
- ❐ Measuring spoons: One teaspoon and 1/4 teaspoon
- ❐ Hammer and nails to secure the rack
- ❐ Felt tip marker

How to Do It

1. Rain gauges measure the amount of rainfall in cubic inches. So your first task is to make a scale for your container that shows how many cubic inches of water are in the container. One cubic inch of water is about 3 1/4 teaspoons, so you can draw the scale on your container by measuring 3 1/4 teaspoons of water to your container, then drawing a short line at the level of the water. If you look closely, the top of the water will seem to be slightly curved and thickened. Draw your line so that it matches the bottom of the curved surface (which is called a meniscus). This line corresponds to a rainfall of one inch.

2. Add another 3 1/4 teaspoons of water to the container and draw another line. The second line corresponds to a rainfall of two inches.

3. Repeat Step 2 until you have at least five marks on the container. This will be enough for most rain events; but you may want to add another line or two, just in case!

4. Find a location for your rain gauge where there is nothing overhead (such as trees or a building roof) that could direct water into or away from your gauge. The edge of a fence away from buildings is often a good spot. Another possibility is to attach your rain gauge to a broomstick driven into the ground in an open area. Be sure to record rainfall soon after a rain event to avoid false readings caused by evaporation.

Empty your gauge after each reading, and you are ready for the next event!

This activity is adapted from "Build Your Own Weather Station" by the Educational Technology Programs Team at the Franklin Institute, Philadelphia, PA (http://www.fi.edu/weather/todo/todo.html).

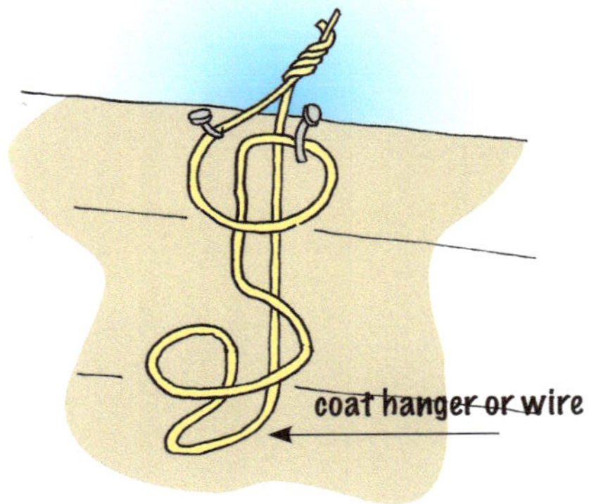

Finished Rain Guage

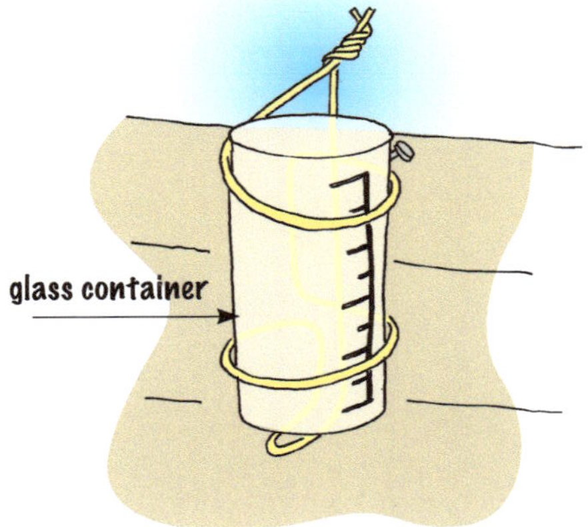

PHOTO AND GRAPHIC CREDITS

Carbon Footprints • Page 43 • CO_2 Levels • Courtesy NOAA
https://gml.noaa.gov/ccgg/trends

Carbon Footprints • Page 44 • CO_2 Calculator • Website courtesy U.S. EPA
www.epa.gov/climatechange/ghgemissions/ind-calculator.html

Cloud Formation • Page 46 • Mammatus clouds • Courtesy NOAA
https://www.nesdis.noaa.gov/about/k-12-education/atmosphere/types-of-clouds

Cloud Types • Pages 49–54 • Ten Basic Clouds • Courtesy NOAA
https://www.noaa.gov/jetstream/clouds/ten-basic-clouds

Cloud Types • Pages 56–57• Cloud Classifications and Characteristics • Courtesy NOAA
https://www.weather.gov/media/lmk/soo/cloudchart.pdf

Global Winds • Page 65 • Global wind pattern • Courtesy NWS NOAA
https://svs.gsfc.nasa.gov/4571

Greenhouse Effect • Page 70 • CO_2 at Mauna Loa • Courtesy NOAA
www.esrl.noaa.gov/gmd/ccgg/trends/#mlo_full

Hurricanes • Page 81 • Cyclone formation regions • Courtesy NWS NOAA
https://www.nhc.noaa.gov/climo

Making Weather Instruments • Page 83 • Rain gauge • Purdue Extension Service

Making Weather Instruments • Page 85 • Wind vane • Purdue Extension Service

Making Weather Instruments • Page 86 • Wind vane construction • Figure 15: Purdue Extension Service

Making Weather Instruments • Page 87 • Anemometer • Purdue Extension Service

Making Weather Instruments • Page 91 • Barometer • Figure 27: Purdue Extension Service

Mini-Tornado • Page 96 • Tornados • Courtesy NOAA,

Out of the Dust • Page 97 • Dust storm • Courtesy of www.dreamstime.com (purchased by designer)

NOTES

LEVEL 3

Weather and Climate Science Education for Grades 9–12

NOTE TO 4-H MEMBER

As you continue your study of weather and climate science, you will be increasingly responsible for collecting your own data and gaining information from sources outside this BOOK. Level 1 introduces basic weather terminology and concepts. Activities focus on understanding the signs of weather around you. Level 2 introduces more complex weather topics such as air pressure, winds, humidity, and fronts.

Level 3 delves even deeper into weather and introduces climate science concepts to help prepare you to fully understand weather and/or study these topics at a college or university. In Level 3 you are encouraged to supplement your learning by consulting knowledgeable people and recent written materials. The study of weather and the earth's climates has evolved significantly, so recent references are preferred.

The key to learning in this, or any project, is to enjoy your studies and to learn at your own pace. We hope this project is just the start of a lifetime enjoyment of understanding the climate you live in and watching the weather!

Safety First!

Ask a parent or other adult to help when you see a safety first icon.

Knowing the Difference

Weather—is the state of the atmosphere at a particular place and time and is influenced by climate and many local factors.

Climate—describes the prevailing or general long-term weather conditions for an area, or for the entire planet.

Climate System—Earth's water and gases that flow or change state as a result of the sun's energy.

Climate—The average weather over time.

Weather—What is happening today and tomorrow.

AIR MASSES AND FRONTS

How are the movements of air masses involved in the formation of fronts and their resulting weather?

RECOMMENDATION: Before beginning, review the Cold Front and Cloud Formation activities from the Level 2 manual.

Air masses have distinct temperature and moisture characteristics. These characteristics develop when air masses spend time over large areas of the earth's surface. As they begin to move, fronts form. Each type of front brings weather that depends on the air masses involved. Knowing the characteristics of air masses and how they move helps meteorologists predict the type of weather that will develop along the resulting fronts.

OBJECTIVE: Explain the development of air masses, the resulting formation of fronts, and the weather those fronts produce.

Weather Tote

Air Masses and Fronts worksheet on pages 125–130

1 Complete the worksheet.

2 Discuss the Chat questions with a parent or other adult helper.

SHARE WHAT HAPPENED: What causes stormy weather in the Midwest?

APPLY: When you see stratus and nimbostratus clouds, what weather would you predict?

GENERALIZE TO YOUR LIFE: How can you use the information you learned to help you understand TV weather reports?

Fly Higher

- Observe weather reports at the NOAA website, www.noaa.gov.
- Observe the fronts approaching your area for several days.
- Keep track of how they move and the kind of weather they cause.

AIR MASSES AND FRONTS WORKSHEET

ACKNOWLEDGEMENT: Some text and graphics for this activity are from the National Oceanic and Atmospheric Administration (NOAA).

Air Mass

An air mass is a large body of air that has relatively uniform temperature and humidity characteristics. The regions where air masses form are called air mass source regions. If air remains over a source region long enough, it will acquire the properties of the surface below. Ideal source regions are generally flat and of uniform composition. Examples include central Canada, Siberia, the northern and southern oceans, and large deserts.

Air masses are classified according to their temperature and moisture characteristics. They are grouped into four categories based on their source region. Air masses that originate in the cold, polar regions are designated with a capital "P" for polar. Air masses that originate in warm, tropical regions are designated with a capital "T" for tropical. Air masses that originate over land are dry and designated with a lowercase "c" for continental. Air masses that originate over water are moist and designated with a lowercase "m" for maritime. In winter, one more type of air mass may form, an extremely cold, dry air mass called "cA," continental arctic. Once formed, air masses can move out of their source regions, bringing cold, warm, wet, or dry conditions to other parts of the world.

Common North American air masses are shown in Figure 31 and described below:

- Maritime tropical (mT) air mass—forms over tropical water. It is warm and moist.
- Continental polar (cP) air mass—forms over a cold land. It is cold and dry.
- Maritime polar (mP) air mass—forms over cold water. It is cold and moist.
- Continental tropical (cT) air mass—forms over warm land. It is warm and dry.

FIGURE 31. **AIR MASSES**

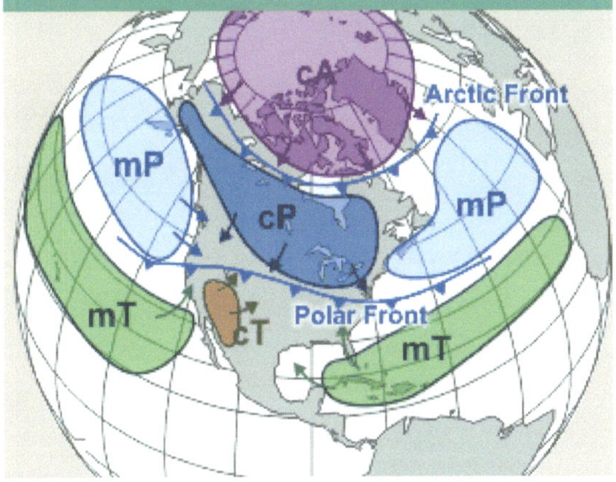

Source: National Oceanic and Atmospheric Administration/ Department of Commerce.

Fronts

A front is the boundary between two air masses. Fronts are classified by the type of air mass (cold or warm) that is replacing the other type of air mass.

The Cold Front

A front is called a cold front if a cold air mass is replacing a warm air mass. The air behind a cold front is colder and typically drier than the air ahead of it, which is generally warm and moist. There is usually a shift in wind direction as the front passes, along with a change in pressure tendency—pressure falls prior to the front arriving and rises after it passes. A cold front has a steep slope, which causes air to be forced upward along its leading edge. This is why a band of showers and/or thunderstorms sometimes lines up along the leading edge of the cold front. Cold fronts are represented on a weather map by a solid blue line with triangles pointing in the direction of its movement.

DID YOU KNOW? The United States has more tornadoes than any other country in the world and four times as many as all of Europe.

Differences in air pressure cause air masses to move. When a cold air mass moves into an area of warmer air, as shown in Figure 32, the denser cold air forces the warmer air to rise at the frontal boundary. Towering cumulus and cumulonimbus clouds form in a cold front. The density difference between the two air masses and the amount of moisture in the warm air determines how severe the weather in the front will be. Cold air is denser than warm air, and dry air is denser than moist air, so cold, dry air is denser than warm, moist air. Density differences between different air masses result in very little mixing of the air.

FIGURE 32. **COLD FRONT**

FIGURE 33. **COLD FRONT AIR MASSES**

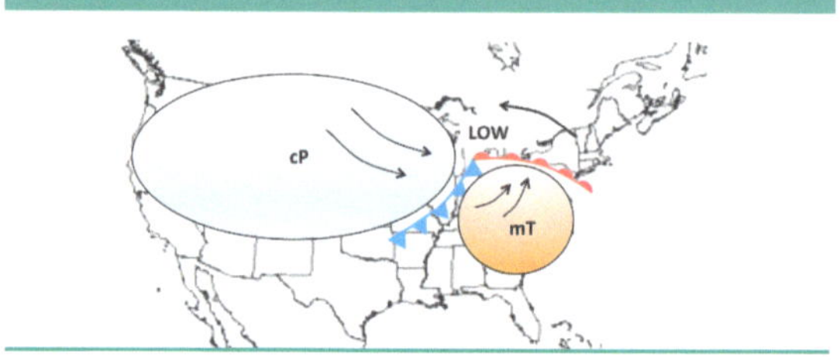

In the Midwest, warm air masses are usually southwest winds, and cold air usually blows in from the northwest, as shown in Figure 33. It is common to experience a change in wind direction from southwest to northwest after a cold front moves through. The cold air mass moves faster and at a slightly different angle than the warm air. As the cold air moves, friction against the ground slows the air near the surface, causing the upper winds to tumble downward at the frontal boundary. A rapid rise in the warm air causes cloud formation; and the moister the warm air, the more clouds develop.

Warm air can be lifted rapidly in advance of a cold front, creating vertical shear winds. Vertical shear winds occur when there is a quick change in vertical direction as the cold front air rises in the warm air and sinks in the cold air.

Weather along a cold front is generally stormy and sometimes severe. In the spring, the continental polar (cP) air masses are extremely dry and cold. These air masses form cold fronts as they move from Canada to the United States and drive into the warm moist air from the Gulf of Mexico. This creates severe cold fronts that can produce damaging storms and tornadoes. Most tornadoes form in the Great Plains, where no landforms disturb airflow. During the winter when the air is cooler in the Midwest, not much moisture is in the air and the cold fronts are not as severe.

The Warm Front

When a warm air mass moves into an area of cooler air, a warm front forms, as shown in Figure 34. Because the warm air is less dense than the cooler air, it cannot push the cooler air out of the way. As a result, the angle of the frontal boundary is not as severe as with a cold front. The warm air climbs slowly over the cooler air, producing layered clouds like stratus and nimbostratus clouds.

FIGURE 34. **WARM FRONT**

FIGURE 35. **WARM FRONT AIR MASSES**

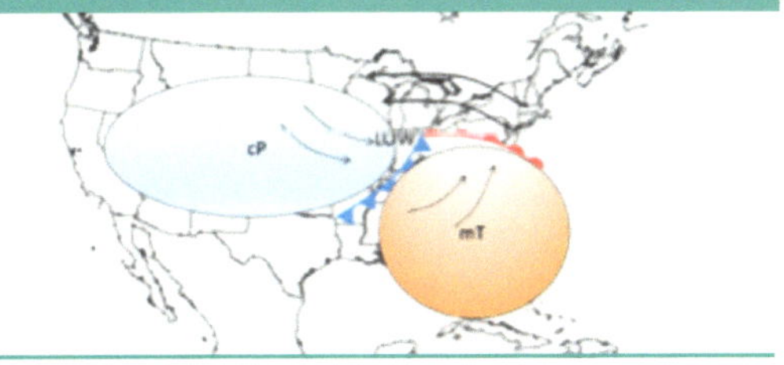

Far in advance of the front, cirrostratus and cirrus clouds form. Warm fronts form clouds that cover a large area and create rain or snow events that can last for days. When we are experiencing warm front-produced weather, the warm air is aloft, and we are in the cooler air. The moist, warm air from the Gulf provides the moisture for the clouds and precipitation. Warm fronts produce heavy snow and ice storms in the Midwest, as the weather produced wraps around to the northern section of the low pressure system, as shown in Figure 35.

Warm fronts produce overcast skies and steady precipitation that may last for long periods. In the winter, warm fronts can produce blizzard conditions; in the spring, they can produce prolonged rain that can cause flooding. The heaviest snow and most severe ice storms occur when the winds are out of the east and northeast as moist air from the Gulf is lifted and pulled around the low-pressure zone.

Other Fronts

Stationary and occluded fronts are variations of cold and warm fronts. A stationary front, as you might guess, is a cold or warm front that comes to a standstill. The warmer air continues to rise over the cooler air, but the position of the front on the earth moves very little. In a stationary front, the cold air normally flattens out, and the resulting front resembles a warm front with stratus and nimbostratus clouds. Stationary fronts form away from the influence of the primary low and are influenced by other factors in the atmosphere. They often lead to flooding. Stationary fronts are represented on a weather map by alternating red and blue lines, with blue triangles and red semicircles facing opposite directions.

FIGURE 36. **COLD FRONT ADVANCING TOWARD A WARM FRONT**

FIGURE 37. **OCCLUDED FRONT**

FIGURE 38. **OCCLUDED FRONT FORMATION**

The boundary between two cold air masses is called an occluded front. Occluded fronts form because cold fronts usually advance faster than warm fronts and catch up to a warm front. A cold front is shown advancing toward a warm front in Figure 36. When the cold front overtakes the warm front, the warm air is lifted between the two cooler air masses. The coldest air mass undercuts the others. Occluded fronts are represented on weather maps by a solid purple line with alternating triangles and semicircles, pointing in the direction of its movement.

The several different types of occluded fronts are based on their interactions due to the density of the two cooler air masses. Figure 37 shows an approaching cold air mass denser than the warm air mass that created a front with another cold air mass. The cold air mass will lift the cool air mass as well as the warm air mass. Occluded fronts form near the center of low pressure systems, as indicated in Figure 38. The weather in an occluded front depends on the location within the front. As the front passes, you can expect stratus and nimbostratus clouds followed by towering cumulus and cumulonimbus clouds.

Questions:

AIR MASSES

Match the air masses to their characteristic (each letter is used twice):

AIR MASS	LETTER	CHARACTERISTIC
A. Maritime Tropical (mT)		Dry and warm
		Moist and cold
B. Continental Tropical (cT)		Moist and warm
		Dry and cold
C. Continental Polar (cP)		Forms over warm land
		Forms over warm water
D. Maritime Polar (mP)		Forms over cold water
		Forms over cold land

THE COLD FRONT

What causes the movement of air masses?

____

____

What causes the warm air ahead of a cold front to rise?

____

____

What kind of clouds usually form in a cold front?

____

What wind direction change usually occurs as a cold front moves through an area?

____

Where do vertical shear winds form?

____

What kind of weather is usually expected in a cold front?

____

THE WARM FRONT

What cloud types are typically associated with a warm front?

____

Which air mass provides the moisture that produces the clouds and precipitation?

____

What severe events can warm fronts produce?

____

What kind of weather is usually expected in a warm front?

____

How does an occluded front form?

____

Within the low, where do occluded fronts form?

____

How do stationary fronts form?

____

Within the low, where do stationary fronts form?

____

ISAAC'S STORM

How might a weather event affect the people living through it?

intro Meteorological observations in the United States were the responsibility of the U.S. Army Signal Corps from 1861 to 1891. Congress, at the request of President Benjamin Harrison, passed an act transferring the meteorological responsibilities of the Signal Corps to the newly created U.S. Weather Bureau in the Department of Agriculture on October 1, 1890.

Isaac's Storm is the story of Isaac Monroe Cline, chief meteorologist at the Galveston, Texas, office of the U.S. Weather Bureau from 1889 to 1901. The book describes the early workings of the Weather Bureau and how a lack of scientific knowledge about hurricanes resulted in the greatest loss of life in U.S. weather history.

OBJECTIVE: Discuss how weather can change people's lives.

Weather Tote

- Book: Larson, Erik. *Isaac's Storm*. New York: Vintage Books, 2000. (ISBN: 0-375-70827-8)

Do it

1 Read *Isaac's Storm* by Erik Larson.

2 Discuss the Chat questions with a parent or other adult helper.

Chat

SHARE WHAT HAPPENED:

- Why was Galveston an important city in Texas?
- What was the general opinion about Galveston's safety from a hurricane in 1900?
- How did Isaac Cline come to be the head of the Galveston weather station?
- Why did the Weather Bureau ignore reports from Havana, Cuba?
- What is the significance of the ring on Isaac's finger in his photograph?
- What part of this book was most interesting to you? Why?

APPLY: How would a hurricane like this be handled differently today?

GENERALIZE TO YOUR LIFE: Can you think of weather events, either positive or negative, that have changed things in your life or the lives of your family members? Explain your answer.

Fly Higher

- Identify another hurricane event. Learn more about it and share your findings with a parent or friend.
- Learn more about the history of NOAA:

  www.noaa.gov/history-of-noaa-1970-1979

  www.weather.gov/timeline

NOTES

MONITORING WEATHER

How can you share your weather knowledge with others?

intro The Community Collaborative Rain, Hail and Snow (CoCoRaHS) network is a group of volunteer backyard weather observers who measure and report precipitation (rain, hail, and snow) events in their local communities. CoCoRaHS volunteers are trained to use low-cost weather measurement tools. They report their weather data online. It is then accessible to researchers, the media, and others interested in weather data. The goal of the CoCoRaHS network is to provide more detailed weather observations than are available from weather stations. For example, CoCoRaHS data is used to help scientists more accurately predict the expected runoff from snowmelt in the Rocky Mountains.

The only requirements to join are enthusiasm for watching and reporting weather conditions and a desire to learn more about how weather can affect our lives. CoCoRaHS volunteers must be trained, make a commitment to the program, obtain an approved rain gauge, and make daily reports.

OBJECTIVE: Attend CoCoRaHS training, become a CoCoRaHS volunteer, and submit precipitation data online for one year.

Weather Tote

- A personal email address
- A CoCoRaHS-approved rain gauge
  - May be obtained through www.cocorahs.org/Content.aspx?page=approved_rain_gauges
  - See Figure 39

FIGURE 39. **RAIN GAUGE**

1 Option 1 (preferred): Contact an Indiana Coordinator at www.cocorahs.org/Content.aspx?page=coord_in, to register for a training near you.

2 Option 2: Visit the CoCoRaHS website: http://www.cocorahs.org/.

- Click on Join CoCoRaHS.
- Complete an application to become a CoCoRaHS participant.
- You will be contacted through email with instructions and logon directions.

3 Collect and report daily precipitation observations.

4 Discuss the Chat questions with a parent, friend, or adult helper.

SHARE WHAT HAPPENED: What did you learn during your CoCoRaHS training?

APPLY: How might the data that you collect be used?

GENERALIZE TO YOUR LIFE: How might the data you collected for CoCoRaHS be used during a major snowstorm?

RESOURCE PEOPLE: Contact one of the following individuals if you have questions about participating in the CoCoRaHS network in Indiana (information downloaded in 2026):

Dr. Beth Hall
Indiana State Climate Office
915 West State Street
West Lafayette IN 47907
(765) 494-8060
bethhall@purdue.edu

Meagan Bird
Indianapolis NWS Office
6900 West Hanna Avenue
Indianapolis IN 46241-9526
(317) 956-0360
meagan.bird@noaa.gov

Earl Breon
Indianapolis NWS Office
6900 West Hanna Avenue
Indianapolis IN 46241-9526
(317) 856-0361
earl.breon@noaa.gov

PRESSURE SYSTEMS—HIGHS AND LOWS

What kind of weather is expected with high and low pressure systems?

intro The sun is constantly adding energy to half of the earth. The half that receives energy depends on the time of day and the tilt of the earth. The amount of energy that earth retains depends on many variables, including land and water surfaces, latitude, and atmospheric conditions.

The earth's rotation influences how global wind patterns develop. Changes in air temperature and humidity cause changes in atmospheric pressure. These factors create areas of different pressures. Air moves in an attempt to establish pressure **equilibrium**.

OBJECTIVE: Explain the structure and development of high and low pressure systems and their effect on our weather.

Weather Tote

- Glass containing ½ cup of cold tap water
- Pie pan
- Warm tap water
- Sugar, ½ cup
- Tablespoon and teaspoon
- Minute timer or stopwatch
- Red food coloring
- Metal or glass pie pan or round cake pan
- Turntable or lazy Susan
- High and Low Pressure Systems worksheet on pages 137–141

Do it

1 Slowly stir 1–2 tablespoons (T) sugar into the glass of cold water until it is dissolved.

- After one minute, stop stirring to see if any sugar is visible.
- Add a teaspoon (t) of sugar, stir, and wait for one minute.
- Continue to add sugar (1 t) until some sugar remains after you stop stirring for a minute.

2 Add a few drops of red food coloring to the sugar water.

3 Fill the pie pan two-thirds full of warm water and place on the turntable.

- Wait a minute to allow the water to become still.

4 Place 1 t of the red sugar water in the center of the pan.

5 Observe what happens to the red sugar water.

6 Place another teaspoon of the red sugar water in the center of the pie pan and slowly turn the turntable counterclockwise for about 15 to 20 seconds.

- Observe what happens to the red sugar water while the turntable is moving.

7 Stop the turntable and let it sit for a minute.

8 Observe what happens to the red sugar water after the pan is still.

9 Complete the High and Low Pressure Systems worksheet.

SHARE WHAT HAPPENED: What was the most interesting part of this activity?

APPLY: Without repeating the experiment above, tell another way you could teach someone your age about the structure and development of high and low pressure systems and their effect on our weather.

GENERALIZE TO YOUR LIFE: How might understanding pressure systems and their resulting weather help you make your weekend plans?

Fly Higher

Observe television or online weather maps for one week to observe high and low pressure areas. Keep track of how they move and the kind of weather that they cause.

NOTES

PRESSURE SYSTEMS—HIGHS AND LOWS WORKSHEET

SUGGESTION: Review the 4-H Weather and Climate Science activity, Earth's Rotation and Global Winds, before beginning this activity.

Development of High and Low Pressure Systems

Changes in air temperature and humidity cause changes in atmospheric pressure. Systems, including air pressure, naturally move toward equilibrium. High and low pressure systems (often referred to as a "high" or "low") follow the principles demonstrated by the pie pan activity. Air in a low-pressure area is forced up as denser air moves toward the center of a low pressure system. In a high pressure system the denser air sinks and pushes lighter air away. Figure 40 shows this basic premise of air sinking in high-pressure areas and rising in the low-pressure areas. This air movement causes surface air to move from a high toward a low.

There are several reasons for differences in air density. Temperature and moisture content affect air density as follows:

- Cooler air is denser than warmer air.
- Dry air is denser than moist air.

FIGURE 40. **AIR RESPONSE TO PRESSURE**

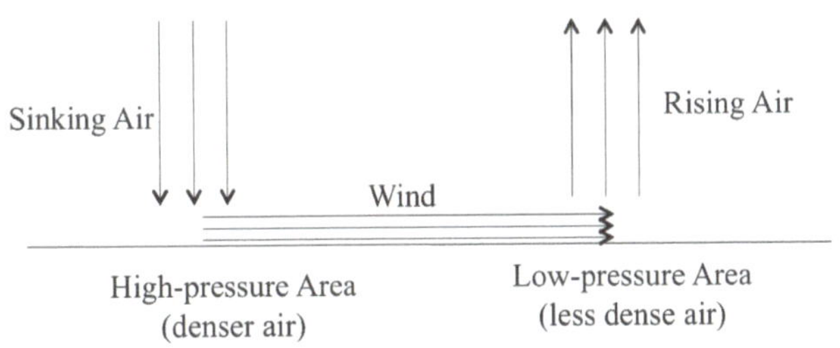

DID YOU KNOW? Galileo saw the Great Red Spot spinning on the planet Jupiter in 1610. It is still spinning more than 400 years later because there is no friction to slow it down.

That's right, moist air is less dense than dry air! This is explained by chemistry: Water molecules (H_2O – atomic weight = 10) are lighter than 99% of the other molecules in the air. For example, N_2 has an atomic weight of 14 and O_2 has an atomic weight of 16). Another way to think of this is to visualize a jar full of marbles (representing nitrogen, 78% of the atmosphere). If you fill a second jar with marbles and marble-sized foam balls (representing water vapor), which jar would be heavier? Of course the jar full of marbles would be heavier than the jar with marbles and foam. Similarly, adding water vapor to the air makes the air less dense because water molecules are light.

Air densities are also affected by swells and depressions in the atmosphere, because air is a fluid and has wave patterns like water in the ocean. Many factors combine to produce areas of sinking air in a high pressure system and areas of rising air in low pressure systems. These factors combine, along with gravity, in an effort to establish a pressure balance (equilibrium) in the atmosphere. Air moves to establish pressure equilibrium. Therefore, air at the surface of the earth moves from high-pressure to lower-pressures areas.

Figure 40 shows movement of air from the point of view of someone standing on the ground. Figure 41 shows air movement from a high in the Great Plains toward a low on the East Coast. Figure 42 shows what should happen with surface winds as air moves from the high toward the low based only on the difference in pressure. But other factors act on the air to prevent the air from simply moving from the high to the low. The rotation of the earth causes air movement across the surface to deflect to the right in the Northern Hemisphere and to the left in the Southern Hemisphere. This is called the Coriolis effect (see Earth's Rotation and Global Winds, Level 2).

FIGURE 41. **AIR MOVEMENT FROM A HIGH TO LOW**

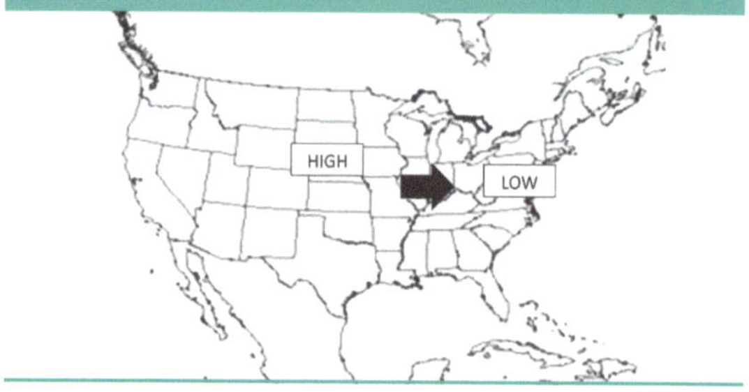

FIGURE 42. **CORIOLIS EFFECT ON AIR MOVEMENT TOWARD A LOW**

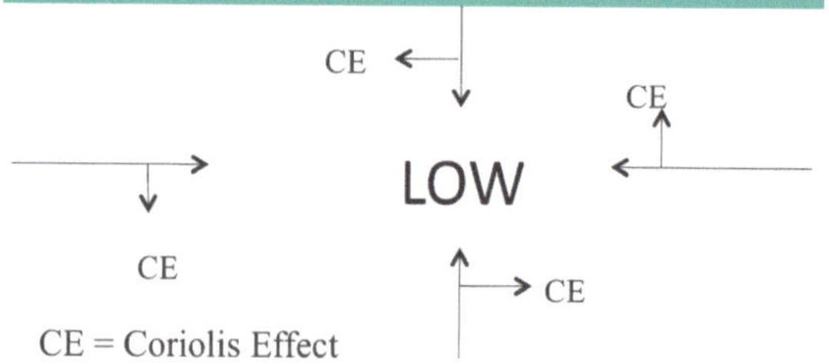

Figure 42 shows a low pressure system with the Coriolis effect (designated by CE) deflecting wind away from the center of the low. This causes the air to circle the low in a counterclockwise direction. Friction, however, causes air near the earth's surface to begin to spiral into the low and works against the Coriolis effect. Friction will eventually cause the storm to dissipate.

Friction and the Coriolis effect influence high pressure systems, too. Dense air sinks at the center of the high, forcing surface air to push out from the center. The Coriolis effect deflects the wind to the right, resulting in a clockwise rotation of wind. Frictional forces at the surface slow the effect of the Coriolis effect, allowing the air to spiral toward the low.

DID YOU KNOW? A centrifugal force impels (pushes) something outward from its center of rotation. For example, if you put 3 cups of water in a small pail and swing the pail on a rope, the water will not drop out when the pail is upside down. Centrifugal force impels the water away from your shoulder, the center of rotation.

DID YOU KNOW?
A car traveling down a highway is not pulled to the right by the Coriolis effect, because the car is in constant contact with the earth. The resulting friction negates the Coriolis effect.

Resulting Weather

High pressure system: Clear and sunny. Rising air may produce fair weather cumulus clouds, which disappear as the sun sets.

Low pressure system: Clouds form in the center of the low and along frontal boundaries.

Figure 44 is a surface map of a typical U.S. weather pattern. Surface maps usually have Hs and Ls to represent high and low pressure systems. On weather maps highs and lows are surrounded by lines called isobars. Isobars are lines of constant pressure connecting locations of equal pressure. When isobars are packed closely together, the pressure is changing rapidly over a small distance. The closer the isobars, the stronger the pressure gradient (the rate of pressure change over a given distance.) Also, notice that in the Northern Hemisphere the wind blows clockwise around a high pressure system and also slightly outward from its center. Around a low pressure system, the wind blows counterclockwise and slightly in toward its center. Because the U.S. is located in the Prevailing Westerlies (see Earth's Rotation and Global Winds, Level 2) these systems move from west to east. Systems like these usually last three to five days.

FIGURE 43. **HIGH AND LOW PRESSURE SYSTEMS**

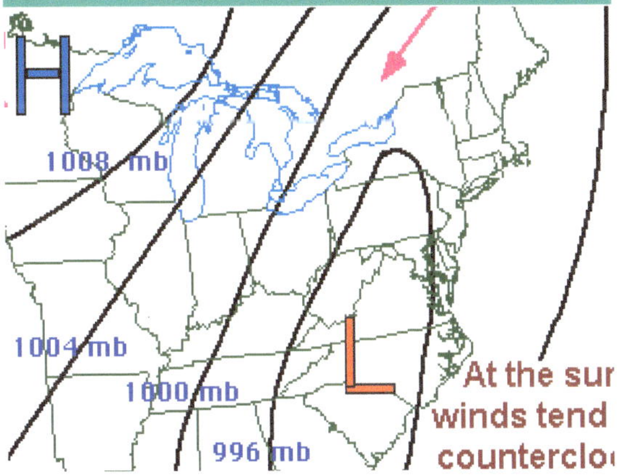

NOAA - Source: National Oceanic and Atmospheric Administration/Department of Commerce.

FIGURE 44. **TYPICAL U.S. WEATHER PATTERN**

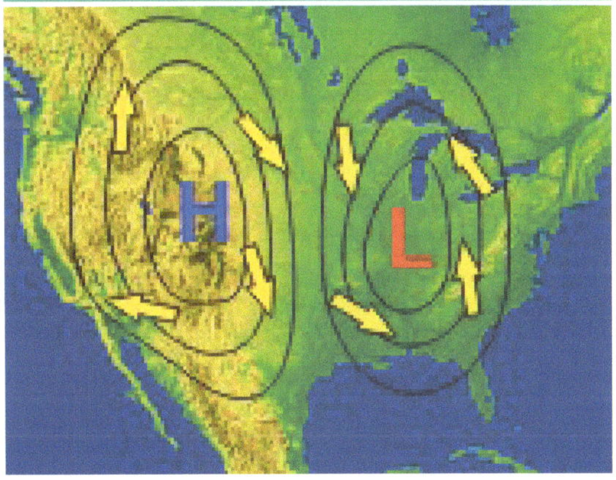

NOAA - Source: National Oceanic and Atmospheric Administration/Department of Commerce.

Questions

PIE PAN ACTIVITY

What did you observe when the red sugar water was dropped into the center of the pan?

What did you observe when you rotated the pan?

What did you observe after the pan was still again?

How does this activity help explain pressure systems?

DEVELOPMENT OF LOW AND HIGH PRESSURE SYSTEMS

Which is denser, warm or cold air?

Which is denser, moist air or dry air?

Which air will form a high pressure system, denser or less dense air?

Which air will form a low pressure system, denser or less dense air?

The Coriolis effect in the Northern Hemisphere causes wind to be deflected which way, to the right or left?

What is the resulting rotation of a high, clockwise or counterclockwise?

What is the resulting rotation of a low, clockwise or counterclockwise?

How is friction involved in the formation of a high or low?

Why do few clouds form in a high?

Why are clouds expected to form in a low?

What kind of weather is expected in a high?

What kind of weather is expected in a low?

WEATHER STATION MODELS

What is the difference between a weather map and a weather station model?

intro Weather information is of interest to everyone. It helps us plan our days and can even be lifesaving during severe weather. The National Oceanic and Atmospheric Administration (NOAA) is the U.S. federal agency responsible for collecting and reporting weather data. NOAA maintains 1,500 **weather stations** to collect weather conditions on land. They also collect ocean data using weather buoys and ships. The ocean data is important because over two-thirds of the earth is covered with water, which has a big influence on our weather. NOAA scientists use that data for many purposes, including determining the global temperature record. Media outlets access NOAA data for their weather reports. You can access your current weather data online at www.noaa.gov by entering your city and state in the search box.

FIGURE 45. **WEATHER MAP**

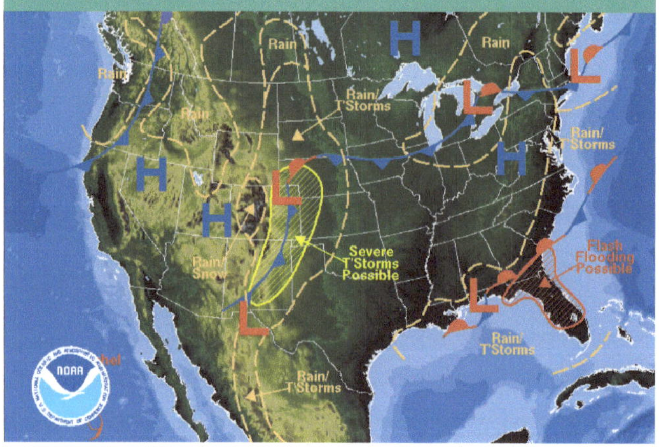

Source: National Oceanic and Atmospheric Administration/Department of Commerce.

Weather maps are used to report current conditions and probable changes in the weather. Meteorologists map weather conditions across the country because weather results from many interrelated variables and generally move in patterns that have been recorded and studied for many years. Weather maps (see Figure 45) show temperature and precipitation patterns and high and low pressure areas. Air movements are studied by examining density and pressure. These movements are called highs and lows. Heavy black lines on a weather map show the fronts and boundaries between air masses. Fronts are important because weather often changes along them. The Air Masses activity (Level 2) introduces many of the symbols used on a weather map.

Meteorologists use **station models** to convey weather conditions at a weather station as succinctly as possible. Information is placed in an organized fashion using symbols to represent the various aspects of weather. The symbols are placed around a circle ("station circle") representing the station where the weather data was collected. An entirely black circle indicates that the skies were completely covered with clouds when the data was collected. Symbols placed around the station circle indicate wind direction, wind velocity, dew point, temperature, visibility, ceiling, air pressure, and types of clouds (if present). The amount of precipitation since the last reading is also recorded. Station models indicate if rain or snow is falling and how the air pressure has changed since the last reading.

FIGURE 46. STATION CIRCLES

SUNNY DAY

CLOUDY DAY

OBJECTIVE: Use weather station model symbols to report weather conditions.

Weather Tote

- Weather Station worksheet on pages 144–146

1 Collect weather data using weather instruments, news reports, and the Internet, as needed.

2 Make a weather station model that shows the weather at your home (or nearby) each day for a week.

- Report the weather at the same time each day.

3 Place each station model on a separate sheet of paper and flip through them.

SHARE WHAT HAPPENED: How many weather symbols can you draw without referring to any reference?

APPLY: How might you use your station model sequence to predict future weather? What additional information do you need?

FIGURE 47. WEATHER STATION MODEL EXAMPLE

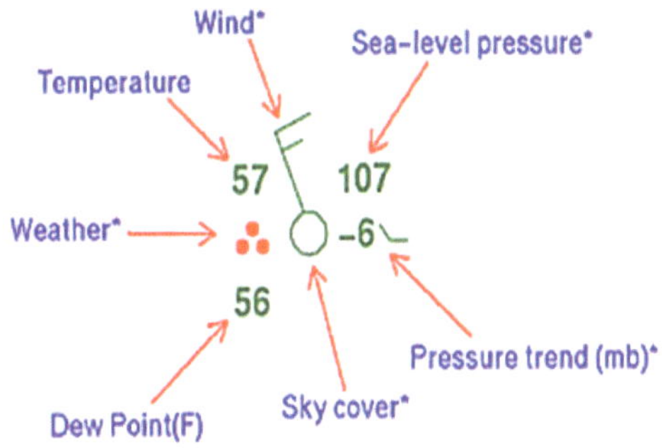

GENERALIZE TO YOUR LIFE:

- Why are symbols and acronyms useful?
- What other symbols can you list?
- What other acronyms can you list?

Fly Higher

- Create a weather map and station models to track a major storm. You must complete this activity when a storm is expected, so planning is required. These websites may be helpful:

  NOAA surface station model map, www.wpc.ncep.noaa.gov/sfc/sfcobs/sfcobs.shtml

  NOAA current severe weather map, www.spc.noaa.gov/products/wwa

  Visit the National Weather Service website, www.weather.gov. Track a storm moving across the U.S.

WEATHER STATION MODELS WORKSHEET

Study the weather symbols given in this worksheet, and create your own weather station model that shows the weather at your house. Use your weather instruments, news reports, or the Internet to collect your weather data.

The symbols for rain, snow, thunder, and fog are shown below. These and other common weather conditions are shown in Figure 48.

FIGURE 48. **COMMON WEATHER CONDITIONS**

Rain	
Drizzle	
A lot of rain (any of the 3 configurations)	
Rain shower	
Heavy rain shower	
Thunder	
Fog	
Snow	*

Wind direction is entered by drawing a line from the direction the wind is coming from to the station circle. North is at the top of the model, south at the bottom, east to the right, and west to the left. Indicate wind speed by drawing lines perpendicular to the wind direction line—a single line for each 10 mph of wind speed, and half a line for each 5 mph of wind speed. These are called feathers and are placed near the end of the wind direction line, away from the station circle. A pennant is used for 50 mph.

FIGURE 49. **WIND SYMBOLS**

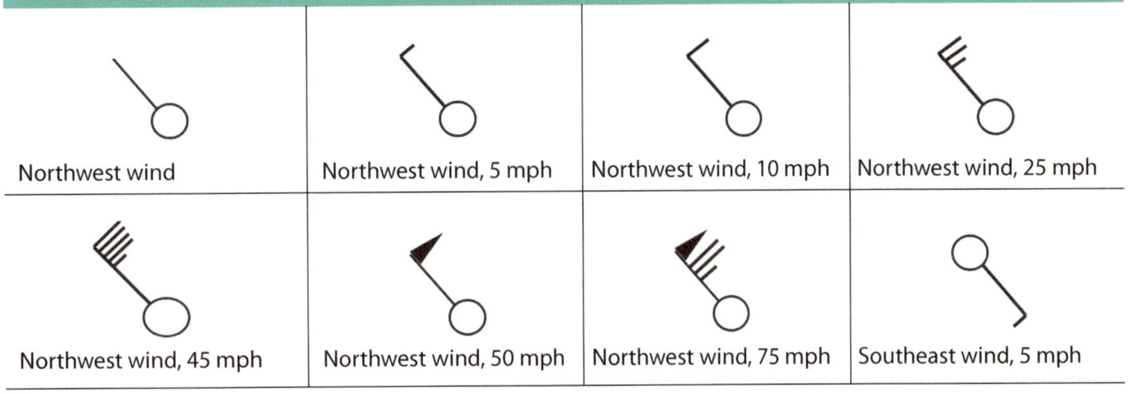

Barometric Pressure—Barometric pressure is plotted in tenths of millibars (mb). The leading 10 or 9 is omitted to simplify the model, so a reading of 1041.0 mb is written as 410. Other examples:

- 103: 1010.3 mb
- 987: 998.7 mb
- 872: 987.2 mb

Barometric Pressure Trend—The pressure trend has two components, a number and a symbol, to indicate how the sea-level pressure has changed during the past three hours. The number provides the three-hour change in tenths of millibars, while the symbol provides a graphic illustration of how this change occurred. Pressure trend symbols are shown in Figure 49.

Optional: NOAA's National Weather Service, Weather Prediction Center provides information to help you read a Weather Station Model at www.wpc.ncep.noaa.gov/html /stationplot.shtml.

FIGURE 50. **COMMON WEATHER SYMBOLS**

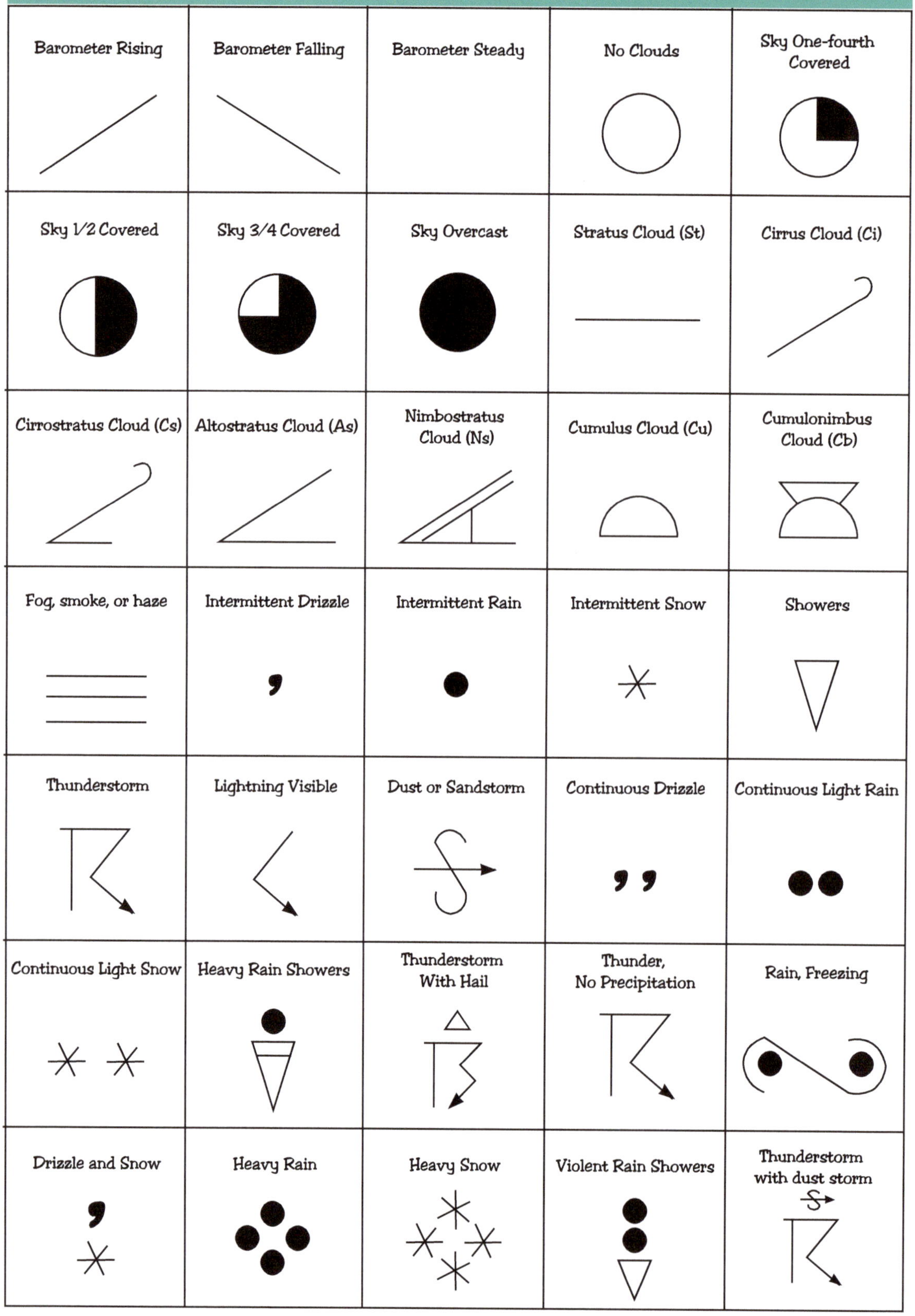

WEATHER IN THE TROPOSPHERE

How does the structure of the atmosphere influence our weather?

SUGGESTION: You will probably understand this activity better if you have completed the following activities: Global Winds (Level 2); Air Masses and Fronts (Level 3); and High and Low Pressure Systems (Level 3).

The earth's atmosphere is divided into four distinct layers based on their temperature and altitude. The layer nearest the surface, where we live, is called the ***troposphere***. The troposphere extends from the surface of the earth to the bottom of the next atmospheric layer, the stratosphere. The troposphere is where all of earth's weather takes place.

The sun's radiation and the earth's atmosphere and surface influence weather dynamics by heating the air in the troposphere differently at different locations. This results in thermal cells, global wind patterns, and jet streams. Understanding how these systems work and how climate changes affect these systems will help you understand global weather patterns.

OBJECTIVE: Understand the dynamics of weather that occurs within the troposphere.

Weather Tote

- Weather in the Troposphere worksheet on pages 149–154

1 Complete the worksheet.

2 Discuss the Chat questions with your adult helper.

SHARE WHAT HAPPENED: In what ways is a jet stream like a river?

APPLY: How could you use what you learned in this activity to improve your study of weather? How might you use it in your everyday life?

GENERALIZE TO YOUR LIFE: How is climate change affecting the weather in the troposphere?

Fly Higher

- Track current jet stream activity and compare it to current weather conditions. The following websites might be useful in your research:
  NOAA Atmosphere information: www.noaa.gov/jetstream/global/jet-stream
- Ten-day forecast of jet stream: www.weatherstreet.com/states/gfsx-300-forecast.htm
- Current winds aloft: www.weather.gov/zan/windsaloft

NOTES

WEATHER IN THE TROPOSPHERE WORKSHEET

Scientists divide the atmosphere into four distinct layers. Each layer is defined by its altitude and the temperature change in the layer. These layers are described below. Processes that affect the weather in the troposphere, the layer nearest the earth's surface, are also described.

Thermosphere

The uppermost layer of the atmosphere (farthest away from the earth) is the **thermosphere.** The thermosphere begins about 55 miles (88 km) above the surface of the earth. It does not have an upper boundary because there is no physical delineation between the atmosphere and space. The oxygen content is very low in the thermosphere, but the oxygen molecules there absorb high-energy radiation and protect us from most of the sun's damaging radiation. This causes the temperatures in the thermosphere to become very high, reaching 2,500°C (4,530°F).

The boundary between the thermosphere and the next layer, the mesosphere, is called the mesopause.

Mesosphere

The **mesosphere** is located between 30 miles (48 km) and 55 miles (88 km) above earth's surface. Very little radiation is absorbed in this layer. Temperature decreases with height in the **mesosphere** because there is little ozone to absorb the radiation.

The boundary separating the mesosphere and the stratosphere, the next layer, is called the stratopause.

Stratosphere

The **stratosphere** is located between seven miles (11 km) and 30 miles (48 km) above the earth's surface. Ozone exists in the stratosphere and absorbs much of the remaining ultraviolet (UV) light from the sun. UV light is damaging to living organisms. The absorption of UV radiation causes the temperatures in the stratosphere to rise.

The air temperature begins to increase with increasing height (temperature inversion) in the **stratosphere**. This warming occurs because ozone in the stratosphere absorbs UV radiation. The ozone also protects life on earth from this dangerous radiation.

The boundary separating the troposphere from the stratosphere is called the tropopause.

Troposphere

The layer at the surface of the earth is the **troposphere**. The troposphere is the layer between the earth's surface and the stratosphere. The top of the troposphere is about 5.5 miles or 30,000 feet, the altitude where most jet aircraft fly.

The troposphere is of particular interest because it contains all of the earth's weather. Sunlight warms the earth's surface, and the surface warms the air above it. Generally, the air becomes colder farther from the earth's surface. Two exceptions to this general rule are a temperature inversion, when air temperature increases with height in a narrow layer; and an isothermal layer, where temperature stays the same with increasing height.

The thickness of the troposphere varies with latitude and the seasons of the year. The troposphere can be more than 20 miles (32 km) thick at the equator and less than four miles (6 km) thick at the North Pole or South Pole. In the U.S., the troposphere is thicker in the summer than in the winter.

FIGURE 51. **ATMOSPHERIC LAYERS**

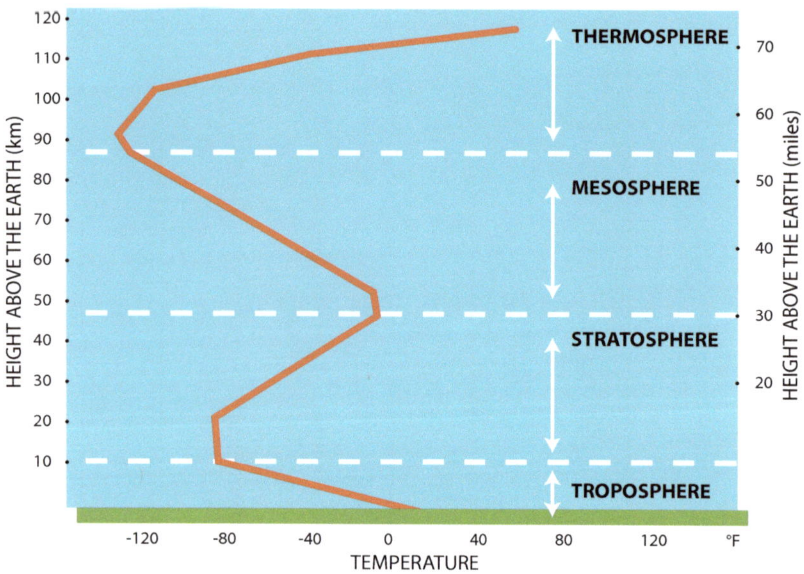

What is the trend of temperature change as the altitude increases?

Thermosphere:____________________ Mesosphere: ____________________

Stratosphere: ____________________ Troposphere: ____________________

Why do you think temperatures increase in the thermosphere?

____

____

Why do you think temperatures increase in the stratosphere?

____

____

Weather in the Troposphere

We live in the troposphere, where almost 90 percent of the earth's atmosphere is located. Air density and temperature change with the height (distance from the earth) in the troposphere. Wind patterns are affected by the rotation of the earth, frictional forces, jet streams, and other influences.

Gravity pulls air molecules toward the earth's surface, so the air is densest at the surface and thins with altitude as oxygen decreases. Mt. Everest has an elevation of almost 5.5 miles (9 km) above sea level. At this elevation, the air is so thin that a breath of air has only 30 percent of the oxygen it has at sea level. This can cause mountain climbers to experience altitude sickness if they are not careful.

Baking can be a challenge at higher elevations. The reduction in air density affects the temperature that causes water to boil. You probably have learned that water boils at 212°F. More accurately, 212°F is the boiling temperature at sea level. The boiling point of water drops about 1°F for each 500-foot increase in elevation. Water boils at about 203°F at 5,000 feet and 195°F at 10,000 feet. Another factor to consider when baking at higher elevations is the effect of the reduced air pressure, which allows leavening gases in baked goods (air, carbon dioxide, and water vapor) to expand more quickly. Baking powder and baking soda amounts must usually be decreased. Liquids and moisture evaporate faster at higher elevations, causing higher concentrations of sugar and fats in the final product than at sea level.

Air temperature declines rapidly in the troposphere as you move away from the earth's surface because the sun warms surfaces on the earth. These surfaces warm the air. For every 1,000-foot rise in elevation, the air is 3.3°F cooler. The sun's energy has very little effect on the temperature of the air in the troposphere; it is either reflected or absorbed by clouds, plants, buildings, and pavement on the earth's surface. The surface warms the air closest to it, so the warmest air in the troposphere is near the earth's surface.

Wind patterns are located within the troposphere (Figure 52). The illustration shows the average position of the wind patterns, but it would be rare to see that *exact* pattern, because many variables affect wind. These cells and winds are constantly fluctuating due to changes in the heating and cooling patterns within the troposphere. The general wind patterns change with the seasons, shifting toward the north during the summer and toward the south during the winter.

FIGURE 52. **WIND PATTERNS**

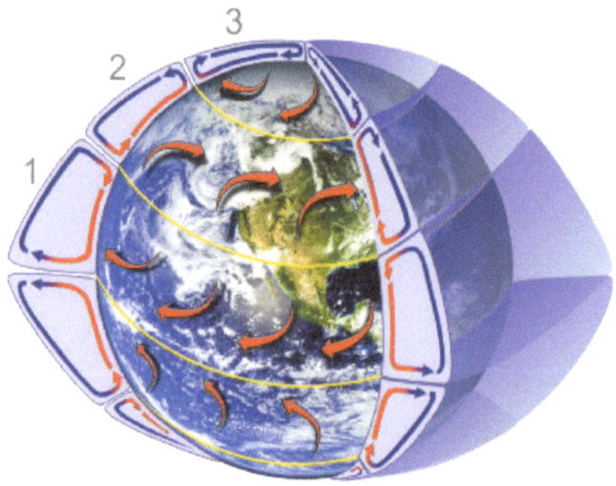

Source: National Oceanic and Atmospheric Administration/Department of Commerce.

Why do mountain climbers sometimes need oxygen tanks?

___

Why is the temperature greatest at the surface of the earth?

___

If it is 65°F at sea level and you are on the top of Mount Everest, what would you expect the air temperature to be, based only on altitude?

___

High- and low-pressure areas develop, and fronts form to produce the weather. Weather systems are influenced by streams of air moving around the earth. These streams of air, which are more accurately described as rivers of air, are called the **jet streams** (see Figure 53). Research on jet streams increased during World War II because of their effect on bomber squadrons in both the Pacific and European theaters. Here is one example:

On November 24, 1944, U.S. bombers flew 30,000–35,000 feet over Tokyo, Japan. The mission was a failure because most of the bombs missed their targets. Technicians realized that the pilots were flying about 90 miles per hour faster than they thought possible. The planes had entered a previously unknown jet stream, which propelled them at higher speeds, just like canoes carried on a swiftly flowing river.

FIGURE 53. **JET STEREAMS**

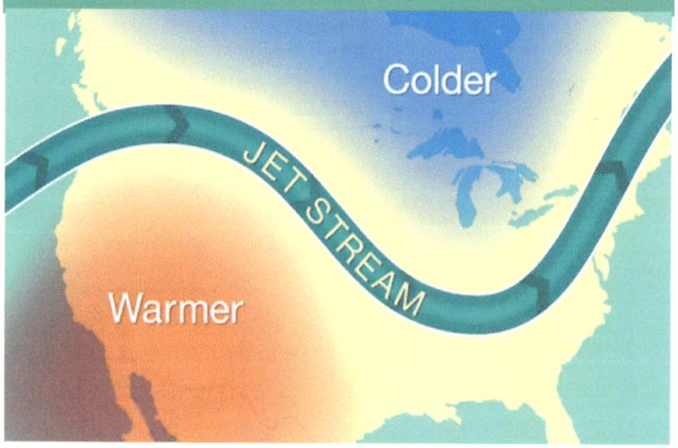

Source: National Oceanic and Atmospheric Administration/Department of Commerce.

Jet streams have speeds of 75 to 150 miles per hour—hurricane- and tornado-scale winds on the surface of the earth. Jet streams maintain these speeds in a west-to-east motion around the globe. Many of the jet streams have been named. The **subtropical jet stream** is located between the Trade Winds and the Westerlies in the area of the Horse Latitudes (30° N and S). The **polar jet stream** is located between the Westerlies and the Polar Easterlies in the area of the Polar Low (60 ° N and S).

A jet stream can be visualized as a tubular river of air with wind speeds greatest in the center. Water in a river flows fastest in the middle, away from the frictional forces that slow flow along the banks and bottom. This is also true for a jet stream. The fastest flow is in the center, farthest from outside frictional forces. This air flow influences the movement of weather systems occurring below, where low pressure areas develop.

How were the jet streams discovered?

____

____

Where is the polar front located?

____

Why is the wind speed highest in the center of a jet stream?

____

____

The polar jet stream shifts north or south depending on the season. During the U.S. summer, when the Northern Hemisphere is tilted toward the sun, the polar jet stream shifts toward the north. In the U.S. winter, when the sun's energy is directed at the Southern Hemisphere, the polar jet stream shifts to the south.

Low-pressure areas commonly travel north in the Midwest during the spring and summer, causing cold fronts and accompanying thunderstorms. That is why thunderstorms and damaging winds are associated with winds from the southwest. In the fall and winter, low-pressure areas travel south, bringing warm front weather that will wrap around the low and bring heavy snow or freezing rain typically associated with east winds. Note, however, that variations can occur.

Normal weather patterns can also be influenced by ocean temperatures. During **El Niño** years, the polar jet stream is forced northward over the U.S. In **La Niña** years, the polar jet stream dips further south. The North Atlantic Ocean can have a profound effect on the polar jet stream over the U.S. When temperatures in the North Atlantic are warmer than usual, and the sea ice in that region does not form, the polar jet stream can become blocked and forced northward.

In normal years, the polar jet stream travels in an easterly direction. When the North Atlantic is warmer than usual, the jet stream is forced northward over the Atlantic, which slows it down. This causes the jet stream over the U.S. to turn southward, bringing cooler temperatures to the southeastern states and heavier snowfall to the northeastern coastal states.

Climate scientists are watching the polar jet stream closely, because abnormal jet stream patterns have increased. Extreme patterns are becoming increasingly frequent, bringing more extreme weather.

The earth is always being warmed near the equator and cooled at the poles. There is a constant tension in the atmosphere as warm air tries to move northward and cold air tries to move southward. This flux results in interesting weather patterns.

Why does the polar jet stream shift northward during our summer?

____

____

Why is cold front weather more severe in the spring and summer in the Midwest?

____

____

Why are warm fronts more severe in the fall and winter in the Midwest?

____

____

How can El Niño and La Niña affect the polar jet stream?

____

____

What will the polar jet stream do when the North Atlantic Ocean remains warm?

____

____

What kind of weather can be expected in the eastern half of the U.S. when there is a polar jet stream blocked by warm North Atlantic temperatures?

____

____

WINDCHILL AND HEAT INDEX

What is a "feels like" temperature?

intro

Knowing the outdoor temperature is helpful when you are deciding what you should wear or how to pack for a day at the beach or on the slopes. Usually you can just read a thermometer to find out the current temperature and have all the information you need. When the temperature is extreme, however, additional information is needed to know how hot or cold it is.

High relative humidity on a hot summer day makes you feel more uncomfortable outside than the temperature would indicate. This is because the high humidity prevents your perspiration from cooling your body and makes the heat feel more intense. It may be more dangerous to be outside, particularly for the very young or very old. Meteorologists report the **heat index**, which is intended to give an accurate idea of what it feels like outdoors.

Wind on a cold winter day makes you feel more uncomfortable outside than the temperature would indicate. This is because the wind pulls heat away from your body at a faster rate. Meteorologists report the **windchill**, which is intended to give an accurate idea of what it feels like outdoors.

Windchill and heat indices are generally reported during times of temperature extremes and are better indicators of health risks than temperatures alone. It is important to know these values during extreme weather so you can dress appropriately and limit outdoor exposure during dangerous weather.

OBJECTIVE: Understand why meteorologists report windchill on cold winter days and heat index on hot, humid summer days.

Note: This activity takes time (est. 2 hours) to allow analog thermometers to give an accurate temperature reading.

Weather Tote

- Two similar thermometers, labeled A and B
- A towel and a metal cookie sheet
- A room that is at a constant temperature
- An electric fan
- The Windchill and Heat Index worksheet on pages 157–162

Do it

Surface and Wind

1 Complete the Windchill and Heat Index Worksheet as you follow the instructions.

2 Record the time you began this activity on the worksheet:

3 Thermometer calibration:

- Place thermometers A and B side by side on a kitchen counter or table.
- Wait 15–20 minutes to allow the thermometers to stabilize.
- If the temperatures do not show the same value, note which one has the lower reading and add the difference to the next several readings taken on that thermometer.

4. Place a towel and a metal cookie tray side by side on the table or kitchen counter.
5. Place thermometer A in the center of the towel and thermometer B in the center of the cookie sheet.
6. Wait 15–20 minutes to be sure the thermometers are stabilized.
7. Place your hand on each surface (towel and cookie sheet).
8. Place the fan so it blows over the thermometers and turn it on.
9. Hold your right hand about 3 inches in front of the fan, keeping your fingers away from the spinning blade. Answer the question in the worksheet.
10. Wait 15–20 minutes to be sure the thermometers are stabilized.
11. Record the time, temperature, units, and surface for each thermometer on the worksheet.

Warming and Cooling

1. Hold the bulb of thermometer A against the palm of your hand until the temperature on the thermometer reaches 90°F.
    - Your readings will be more accurate if you recruit someone else to watch the time, so you can concentrate on the change in temperature.
2. Hold the thermometer away from your body and time how many seconds it takes for the temperature to fall to 80°F.
3. Record the time on the Heating and Cooling Results Chart.
4. Repeat steps 1 and 2 four more times and record the time it took to lose 10°F (resulting in a reading of 80°F) in the chart.
5. Repeat this activity with the thermometer held in front of the fan during cooling.
    - Hold the thermometer in your palm to warm it to 90°F.
    - Hold the thermometer in front of the fan and time how many seconds it takes to lose 10°F.
6. Calculate and record the averages of the five times and record the average in your data chart.
7. Read the What it Feels Like section of the worksheet and answer the questions.

SHARE WHAT HAPPENED: How close were your five readings in the Heating and Cooling Results Chart?

APPLY:

- Why might a room feel cool when one or two people are in it but warm when 100 people are in it, even though the air-conditioning setting has not changed?
- What did you learn from this activity that can make you safer?

GENERALIZE TO YOUR LIFE: Why is it important for parents to understand the windchill and heat index temperatures?

WINDCHILL AND HEAT INDEX WORKSHEET

Surface and Wind

Time: ______________________________

Do you think a pair of thermometers placed near each other, but on two different surfaces, will register the same temperature?

- Yes
- No

Time: ______________________________

Thermometer	Temperature	Units (°F or °C)	Surface
A			
B			

Do the surface temperatures feel the same on the towel and cookie sheet?

- Yes
- No

Which surface feels cooler? ______________________________

How does the air blowing on your right hand feel compared to your left hand, which is not in front of the fan?

- Cooler
- Same
- Warmer

Time after fan blew over thermometers: ______________________

Thermometer	Temperature	Units (°F or °C)	Surface
A			
B			

Did the fan change the temperature of a thermometer that is at room temperature?

- Yes
- No

HEATING AND COOLING RESULTS CHART:

Time for a 10°F drop in temperature ________________

Repetition (Rep)	No Wind Time (sec.)	With Wind Time (sec.)
1		
2		
3		
4		
5		
Average		

How much faster did the thermometer cool when the fan was blowing on it? Why?

____

____

What it Feels Like

You probably found that, although the towel and cookie sheet are at the same temperature, the metal cookie sheet felt colder to your hand than the towel. And you probably found that the air from the fan feels colder to your hand than the surrounding air, but did not lower the thermometer temperature.

Why do you think that is?

____

____

The Reason

Your hand is warmer than the cookie sheet, the towel, and the air at room temperature. Because nature tries to reach equilibrium, more heat is flowing out of your hand than is flowing in. But the temperatures will never equalize because mammals generate heat to maintain a constant body temperature.

The metal cookie sheet feels colder than the towel because metal conducts (transfers) heat better than a towel. Therefore heat from your hand moves to the cookie sheet faster than it does to the towel. So the cookie sheet feels cooler than the towel.

Similarly, the fan creates moving air that removes heat from your right hand faster than the motionless air next to your left hand. Therefore both the temperature of the cookie sheet and the air from the fan feel cooler, but they are really the same temperature. If your hand was the same temperature as the cookie sheet or the air from the fan, you would not feel this difference.

Scientifically speaking, why should you use a towel or hot pad to grab a hot cookie sheet?

Windchill

The temperature of the air feels differently on your skin when a brisk wind is blowing. Just as you saw with the thermometer-cooling test, moving air cools warm objects faster than standing air as heat is moved away from your skin. Therefore wind will chill objects faster than still air. The greater the wind speed against a warmer object, the faster it removes heat and the greater the chill.

Windchill temperature is not an actual temperature; it is a representation of the temperature standing air would have on your skin. Windchill makes air feel colder than it actually is. Scientists have developed the Windchill Chart to show the relationship of temperature and wind speed. The greater the wind speed, the cooler the "feels like" temperature. Even though the windchill temperature is below the freezing point of water, water will not freeze, nor will your skin freeze, as long as the actual temperature is above the freezing point.

FIGURE 54. **WINDCHILL CHART**

Temperature (°F)

Wind (mph) / Calm	40	35	30	25	20	15	10	5	0	-5	-10	-15	-20	-25	-30	-35	-40	-45
5	36	31	25	19	13	7	1	-5	-11	-16	-22	-28	-34	-40	-46	-52	-57	-63
10	34	27	21	15	9	3	-4	-10	-16	-22	-28	-35	-41	-47	-53	-59	-66	-72
15	32	25	19	13	6	0	-7	-13	-19	-26	-32	-39	-45	-51	-58	-64	-71	-77
20	30	24	17	11	4	-2	-9	-15	-22	-29	-35	-42	-48	-55	-61	-68	-74	-81
25	29	23	16	9	3	-4	-11	-17	-24	-31	-37	-44	-51	-58	-64	-71	-78	-84
30	28	22	15	8	1	-5	-12	-19	-26	-33	-39	-46	-53	-60	-67	-73	-80	-87
35	28	21	14	7	0	-7	-14	-21	-27	-34	-41	-48	-55	-62	-69	-76	-82	-89
40	27	20	13	6	-1	-8	-15	-22	-29	-36	-43	-50	-57	-64	-71	-78	-84	-91
45	26	19	12	5	-2	-9	-16	-23	-30	-37	-44	-51	-58	-65	-72	-79	-86	-93
50	26	19	12	4	-3	-10	-17	-24	-31	-38	-45	-52	-60	-67	-74	-81	-88	-95
55	25	18	11	4	-3	-11	-18	-25	-32	-39	-46	-54	-61	-68	-75	-82	-89	-97
60	25	17	10	3	-4	-11	-19	-26	-33	-40	-48	-55	-62	-69	-76	-84	-91	-98

Frostbite Times: 30 minutes; 10 minutes; 5 minutes

$$\text{Wind Chill (°F)} = 35.74 + 0.6215T - 35.75(V^{0.16}) + 0.4275T(V^{0.16})$$

Where, T= Air Temperature (°F) V= Wind Speed (mph) *Effective 11/01/01*

Source: National Oceanic and Atmospheric Administration/Department of Commerce.

Use the Windchill Chart and answer the following questions:

What is the windchill at 40°F with a wind speed of 5 mph? ______________________

What is the windchill at 40°F with a wind speed of 25 mph? ______________________

Will water freeze when the air temperature is 35°F with a windchill that is 28°F?

- Yes
- No

Heat Index

When it is hot outside and the relative humidity is high, the temperature of the air feels differently on your skin than it does when the relative humidity is low. You perspire more. That is your body trying to get rid of excess heat to maintain a normal body temperature of 98.6°F. Your skin is cooled when perspiration evaporates because evaporation carries latent heat energy away from the surface of your skin.

Humid air holds a lot of moisture. Evaporation does not occur as rapidly as in dry air, so cooling does not occur as effectively.

The heat index is a "feels like" indication, not an actual temperature. It is a representation of the temperature standing air would have on your skin. The heat index tells us what the air would feel like if you were in the shade and the air was not moving. Scientists have developed the Heat Index Chart to show the relationship of temperature and relative humidity. The greater the relative humidity, the hotter the "feels like" temperature.

FIGURE 55. **HEAT INDEX CHART**

Temperature (°F)

Relative Humidity (%)	80	82	84	86	88	90	92	94	96	98	100	102	104	106	108	110
40	80	81	83	85	88	91	94	97	101	105	109	114	119	124	130	136
45	80	82	84	87	89	93	96	100	104	109	114	119	124	130	137	
50	81	83	85	88	91	95	99	103	108	113	118	124	131	137		
55	81	84	86	89	93	97	101	106	112	117	124	130	137			
60	82	84	88	91	95	100	105	110	116	123	129	137				
65	82	85	89	93	98	103	108	114	121	128	136					
70	83	86	90	95	100	105	112	119	126	134						
75	84	88	92	97	103	109	116	124	132							
80	84	89	94	100	106	113	121	129								
85	85	90	96	102	110	117	126	135								
90	86	91	98	105	113	122	131									
95	86	93	100	108	117	127										
100	87	95	103	112	121	132										

Likelihood of Heat Disorders with Prolonged Exposure or Strenuous Activity

- Caution
- Extreme Caution
- Danger
- Extreme Danger

Source: National Oceanic and Atmospheric Administration/Department of Commerce.

Use the Heat Index Chart to answer the following questions:

What is the heat index at 82°F with 50 percent relative humidity? ______________

What is the heat index at 82°F with 85 percent relative humidity? ______________

In Arizona the temperature is 88°F with a relative humidity of 45 percent. What is the heat index? What is the danger level for strenuous activity?

____

In Indiana on the same day the temperature is 88°F with a relative humidity of 75 percent. What is the heat index? What is the danger level for strenuous activity?

____

People say that Arizona has a dry heat. What do you think this means?

____

____

Why is it important to drink plenty of water on days with a high heat index?

____

____

Note from NOAA: IMPORTANT: Because heat index values were devised for shady, light wind conditions, exposure to full sun can increase heat index values by up to 15°F. Strong winds, particularly with very hot, dry air, can be extremely hazardous.

The Heat Index Chart shaded zone above 105°F (orange or red) shows a level that may cause increasingly severe heat disorders with continued exposure or physical activity.

Safety First!

"Feels like" events require common sense. In windchill situations, minimize the time people and domestic animals spend outdoors. Cover your exposed skin as much as you can. The warmer the air gets above normal skin temperature, the more dangerous wind is, because it adds heat to your body. When the heat index is high, stay in the shade and use a fan if you don't have air-conditioning.

What can you do to protect yourself in severe windchill situations?

What can you do to protect yourself in severe heat situations?

CLIMATE AND CLIMOGRAPHS

What can you learn by comparing climographs of different locations?

Many elements determine climate, including latitude, elevation, ocean currents, and position on a land mass. These factors influence temperature and precipitation, the major components of a region's climate. For instance, ocean currents produce a warmer climate in some coastal towns than inland towns at the same latitude, and cities at high altitudes generally are dryer than cities at lower altitudes.

A **climograph** s a visual representation of a location's temperature and precipitation. Climographs give us an idea of the type of weather expected at a specific location.

OBJECTIVE: Read and construct climographs.

Weather Tote

- Climate and Climographs worksheet on pages 164–169

1 Complete the worksheet.

2 Discuss the Chat questions with a parent or other adult helper.

SHARE WHAT HAPPENED: What does a climograph show?

APPLY: What can you learn by comparing climographs from different cities?

GENERALIZE TO YOUR LIFE: How might the climographs that you studied in this activity change over the next 50 years?

Fly Higher

- You can view the 1991–2020 climate normals (climographs) at https://www.climate.gov/maps-data/dataset/1991-2020-climate-normals-climographs. Scroll down to "How do I used the site?" for instructions on how to access the data.
- Collect pictures from the different international cities studied in the worksheet. Make a display with the climographs and pictures that reflect the climate at different times of the year.

CLIMATE AND CLIMOGRAPHS WORKSHEET

A climograph is a graphical representation of the temperature and precipitation at a specific location. The average monthly temperature and precipitation for Indianapolis is given below. The climograph for Indianapolis is shown in Figure 57.

The climograph of this data is shown in Figure 56.

FIGURE 56. **INDIANAPOLIS DATA**

	Jan	Feb	Mar	Apr	May	June	July	Aug	Sept	Oct	Nov	Dec
Average Temp (°F)	28.1	31.1	42.2	53.0	62.7	72.0	75.4	74.2	66.9	55.0	43.6	31.6
Precipitation (inches)	2.7	2.3	3.6	3.8	5.1	4.3	4.6	3.1	3.1	3.1	3.2	3.7

FIGURE 57. **INDIANAPOLIS CLIMOGRAPH**

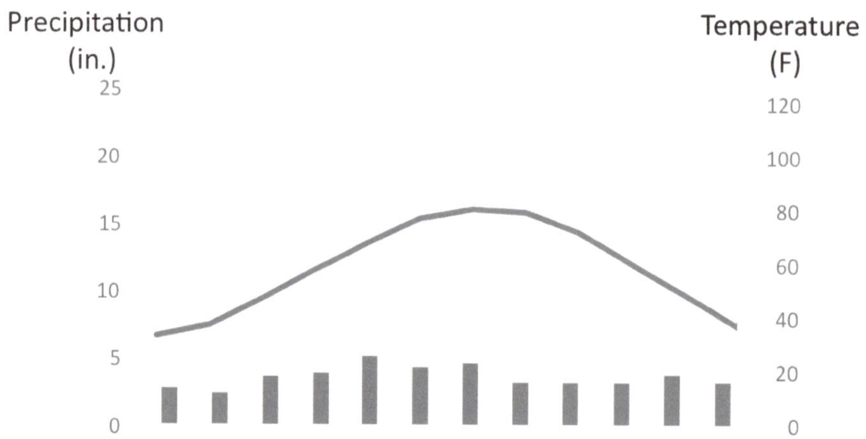

What is the warmest month in Indianapolis, according to the climograph?

____

Which is the wettest month?

____

Which months would probably be best for growing crops?

___

How does the Indianapolis growing season compare to the growing season where you live?

___

___

___

During which months is the average temperature in Indianapolis below freezing (32°F)?

___

Do you think they sell more long underwear or bathing suits annually in Indianapolis?

___

Why do you think so?

___

___

___

Average temperature and precipitation data is given for six U.S. cities below. Use this data to create a climograph for each city and answer the questions.

FIGURE 58. **CITY A**

	Jan	Feb	Mar	Apr	May	June	July	Aug	Sept	Oct	Nov	Dec
Temp (°F)	29.0	31.7	38.3	48.1	57.9	67.7	73.4	72.1	64.9	54.0	44.7	34.7
Precip (in.)	3.4	3.3	4.3	3.7	3.5	3.7	3.3	3.4	3.9	3.9	4.0	3.8

FIGURE 59. **CITY B**

	Jan	Feb	Mar	Apr	May	June	July	Aug	Sept	Oct	Nov	Dec
Temp (°F)	56.4	59.7	65.2	72.7	82.1	90.8	94.8	93.6	88.4	76.7	64.1	55.4
Precip (in.)	0.9	0.9	1.0	0.3	0.1	0.0	1.1	1.0	0.6	0.6	0.7	0.9

FIGURE 60. CITY C

	Jan	Feb	Mar	Apr	May	June	July	Aug	Sept	Oct	Nov	Dec
Temp (ºF)	68.2	70.2	72.6	75.8	79.9	82.7	84.1	84.2	82.9	79.9	74.9	70.5
Precip (in.)	1.6	2.3	3.0	3.1	5.3	9.7	6.5	8.9	9.9	6.3	3.3	2.0

FIGURE 61. CITY D

	Jan	Feb	Mar	Apr	May	June	July	Aug	Sept	Oct	Nov	Dec
Temp (ºF)	23.8	27.7	37.9	48.9	59.1	68.9	74.0	72.4	64.6	52.5	40.3	27.7
Precip (in.)	1.7	1.8	2.5	3.4	3.7	3.5	3.7	4.9	3.2	3.2	3.2	2.3

FIGURE 62. CITY E

	Jan	Feb	Mar	Apr	May	June	July	Aug	Sept	Oct	Nov	Dec
Temp (ºF)	41.4	43.8	48.2	52.3	58.3	63.6	69.2	69.5	64.5	54.9	46.6	40.4
Precip (in.)	4.9	3.7	3.7	2.7	2.5	1.7	0.7	0.7	1.5	3.0	5.6	5.5

FIGURE 63. CITY F

	Jan	Feb	Mar	Apr	May	June	July	Aug	Sept	Oct	Nov	Dec
Temp (ºF)	10.2	15.1	25.9	39.6	51.4	60.1	65.8	64.3	55.6	43.2	28.2	14.8
Precip (in.)	1.0	0.8	1.5	2.4	3.2	4.2	3.9	3.7	4.1	2.9	2.1	1.2

Which city (A–F) has the most precipitation?

___

Which city (A–F) has the least precipitation?

___

Which city (A–F) has the highest temperatures?

___

Which city (A–F) has the lowest temperatures?

____

Which city (A–F) has the most distinct change in seasons?

____

Which city (A–F) has the least changes in seasons?

____

Circle what you would expect to have most of in your closet to be prepared for the weather in each city. (If you cannot decide, circle both.)

CITY	MOST	MOST	MOST	MOST
A	shorts or pants	sandals or shoes	raincoats or sunscreen	heavy coats or light jackets
B	shorts or pants	sandals or shoes	raincoats or sunscreen	heavy coats or light jackets
C	shorts or pants	sandals or shoes	raincoats or sunscreen	heavy coats or light jackets
D	shorts or pants	sandals or shoes	raincoats or sunscreen	heavy coats or light jackets
E	shorts or pants	sandals or shoes	raincoats or sunscreen	heavy coats or light jackets
F	shorts or pants	sandals or shoes	raincoats or sunscreen	heavy coats or light jackets

The temperature and precipitation data is for the cities of Boston, Chicago, Miami, Phoenix, Portland, and Duluth. Indicate the city and state you think the climograph data given on pages 168 and 169 (cities A–F) corresponds to.

	City	State
A	____________________	____________________
B	____________________	____________________
C	____________________	____________________
D	____________________	____________________
E	____________________	____________________
F	____________________	____________________

International Climographs

The international climographs show the cities listed below.

- Beijing, China
- Brasilia, Brazil
- London, England
- Moscow, Russia
- Niamey, Niger
- Sydney, Australia
- Tokyo, Japan
- Toronto, Canada

The line shows temperatures in °F, and the bar graphs show precipitation in inches. Both are shown on the same scale: 0 to 120. Guess which city and country each climograph represents. Check your answers with your facilitator, or check for average monthly temperatures online.

FIGURE 64. **INTERNATIONAL CLIMOGRAPHS**

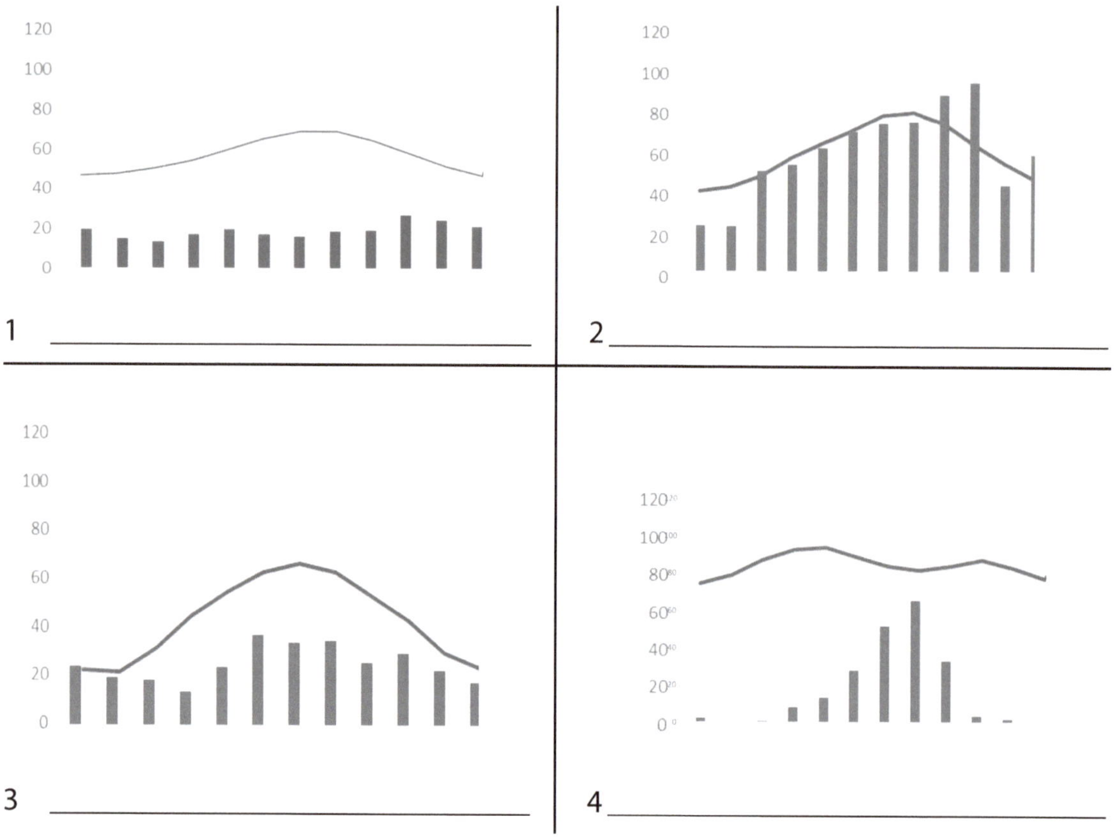

FIGURE 65. **INTERNATIONAL CLIMOGRAPHS**

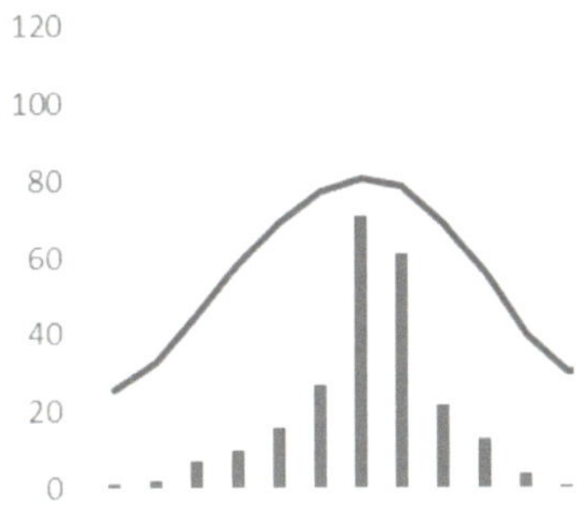

5 ______________________________

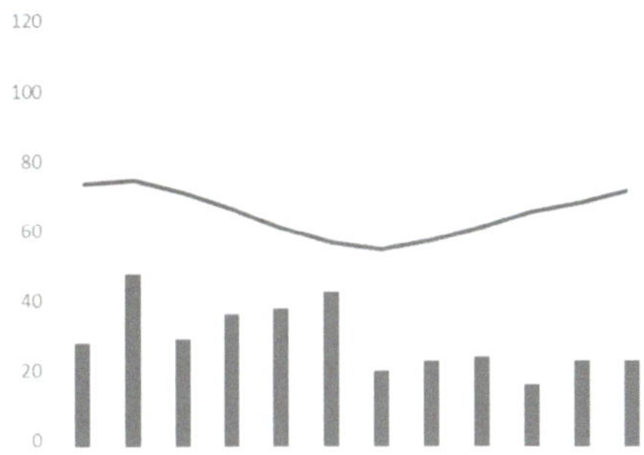

6 ______________________________

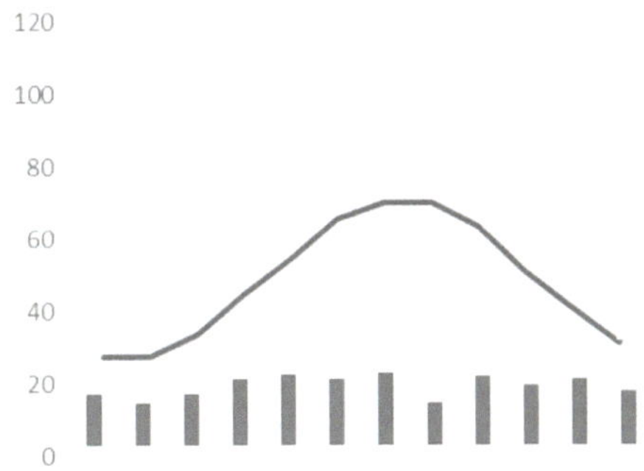

7 ______________________________

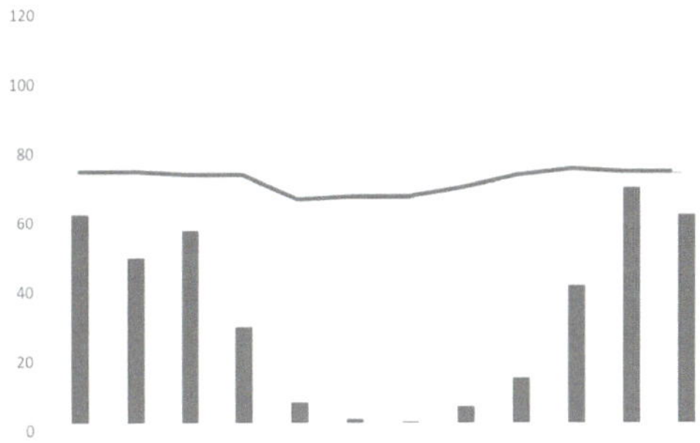

8 ______________________________

DROUGHT MONITORING

How do drought conditions affect people and the environment?

intro Drought can have devastating effects on an area's environment, economy, and people. Droughts are often accompanied by unusually hot temperatures that can be stressful and a health concern for people, particularly the elderly and very young. Droughts affect plant growth—crops, gardens, and trees. They also reduce water levels in streams, lakes, ponds, and groundwater. The environmental and economic impact of a drought can last for years.

Drought occurs when there is not enough rainfall to sustain normal conditions. Normal conditions are based on 30-year averages for the area. Climate data is calculated and updated at the end of each decade at weather stations across the country. Weather conditions in 2015 would be compared to the average rainfall from 1980–2010 to determine if drought conditions exist in a particular area. If the rainfall is significantly lower than the average, the area is considered drought-stricken.

Droughts are big news and often talked about when a community is living through one. But at other times we tend to forget the stress of not having our usual precipitation, particularly in the Midwest, which is fortunate to generally have reliable precipitation. Other areas of the United States and the world are not so fortunate, and drought can be a constant threat. So while you may not currently be experiencing drought conditions, it is likely that drought is a major concern somewhere in the world as you read this. In this activity, you will investigate a recent U.S. drought and its impact.

OBJECTIVE: Research the agricultural, environmental, economic, and social effects of drought on an area.

Weather Tote:

- Internet access for data using the NOAA drought monitor, www.ncdc.noaa.gov/sotc/drought webpage
- Graph paper or computer software to make graphs
- Access to news reports concerning drought effects on a region or state

DO it

1. Use the NOAA drought monitor data, www.ncdc.noaa.gov/sotc/drought, to find a recent drought episode in a region or state.
    - Collect temperature and precipitation data each week for two months.

2 Use line and/or bar graphs to make a **climograph** showing the temperature and precipitation data you collected (see Figure 66).

- Show the dates of data collection on the x-axis.
- Show the temperature and precipitation data on two y-axes as seen for Indianapolis, IN, in the figure.

3 Research the effects of that drought on that region through news reports.

- Include how the drought influenced agriculture, the environment, local economics, and social activities.
- Keep a list of all resources that you use.

4 Create a report or display showing the drought data and your research results on the effects.

5 Discuss the Chat questions with a parent, friend, or adult helper.

SHARE WHAT HAPPENED: What location and year did you study?

APPLY: How might climate change affect drought monitoring?

GENERALIZE TO YOUR LIFE: How could an increase in the incidence of droughts affect your landscaping plans for the first home you purchase?

Fly Higher

Use the NOAA drought website, www.ncdc.noaa.gov/sotc/drought, to study previous droughts, wildfires, tornadoes, hurricanes, snow, and ice in the U.S. or globally.

FIGURE 66. **CLIMOGRAPH FOR INDIANAPOLIS, INDIANA**

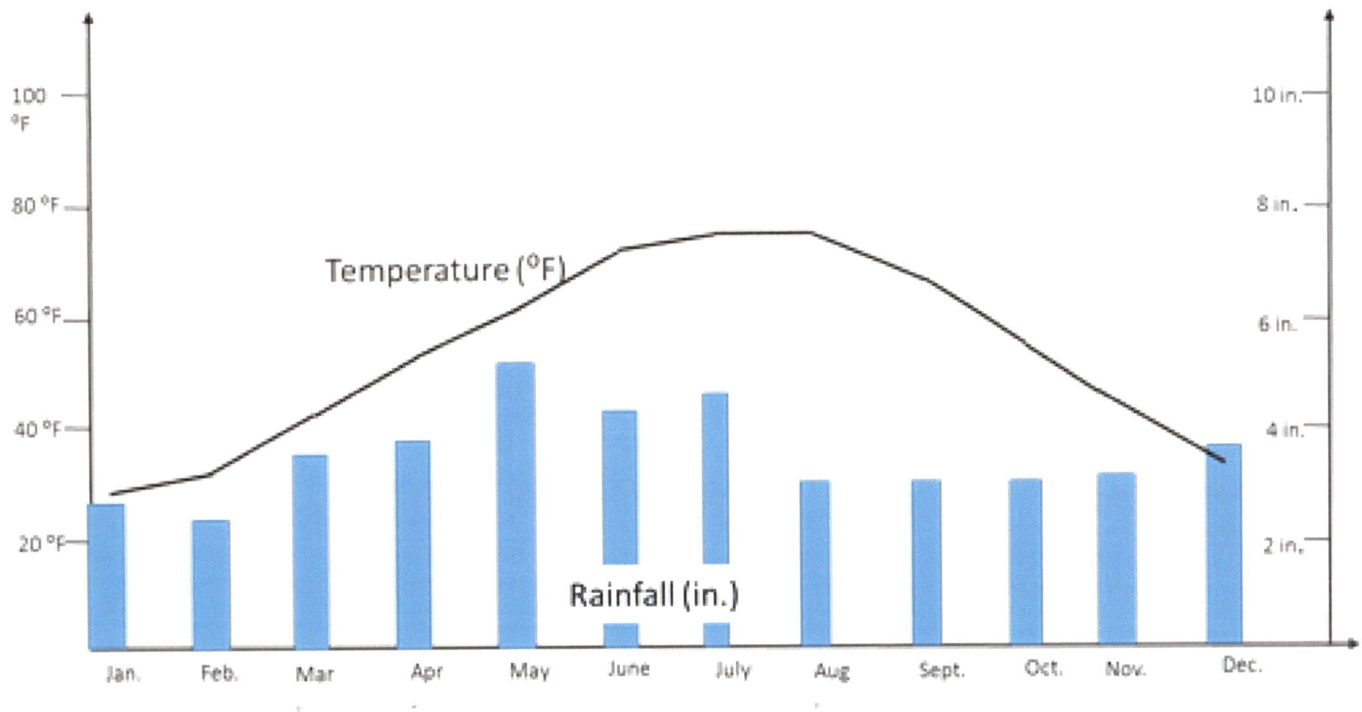

FIGURE 67. PALMER Z INDEX SHORT-TERM CONDITIONS

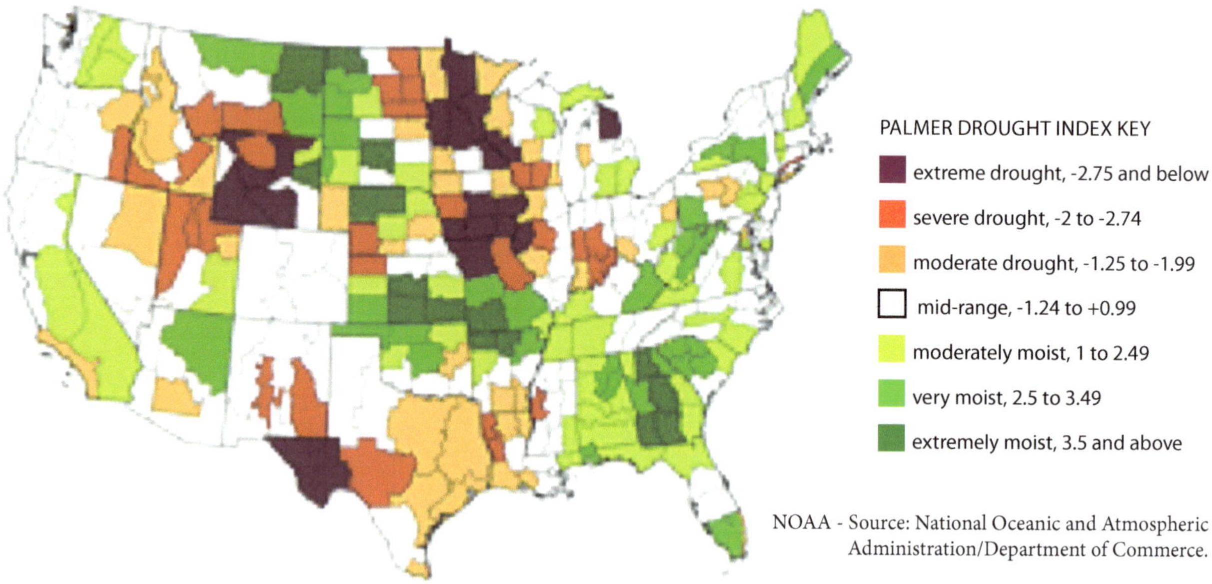

NOAA - Source: National Oceanic and Atmospheric Administration/Department of Commerce.

NOTES

ENERGY IN THE ATMOSPHERE

How does energy from the sun warm the atmosphere?

intro

Radiation from the sun provides the energy that determines earth's weather. This energy generates weather patterns and influences conditions that make life possible. Radiant energy from the sun is absorbed and distributed to achieve the proper balance of energy on earth. You will learn about radiation energy and two other types of energy, convection and conduction.

OBJECTIVE: Describe how radiation, conduction, and convection transfer energy throughout the earth system.

Weather Tote

- Candle
- Paper clip
- Cup or bowl of water
- Pliers (needle-nose, if possible)
- Energy in the Atmosphere worksheet on pages 175–178

Do it

Radiation

1 Light the candle.

2 Hold the palm of your hand about 2 inches from the side of—***not above***—the flame.

3 Move your hand away from the flame.

4 Measure the distance between your hand and the flame when you can no longer feel any heat.

5 Answer the questions on the worksheet and read the radiation information.

Conduction

6 Hold a piece of scrap paper in one hand.

7 Use the pliers to hold a plain metal (uncoated) paper clip in the candle's flame for 15 seconds.

8 Remove the paper clip from the flame and immediately touch it to the piece of paper.

9 Dip the end of the paper clip in the cup of water to cool.

10 Answer the questions on the worksheet and read the conduction information.

Convection

11 Place your hand 3 feet above the flame.

12 Slowly lower your hand until you feel heat from the candle.

13 Measure the distance between your hand and the flame when you can no longer feel any heat.

14 Answer the questions on the worksheet and read the convection information.

15 Discuss the Chat questions with a parent or other adult helper.

Energy from the sun

16 Read the information in the worksheet and answer the questions.

Chat

SHARE WHAT HAPPENED: Name the three main types of energy transfer.

APPLY: Why can it be helpful to wear light-colored clothes on a hot, sunny day and dark-colored clothes on a cold, sunny day?

GENERALIZE TO YOUR LIFE: How might the albedo, or reflecting power, of an area affect your comfort?

NOTES

ENERGY IN THE ATMOSPHERE WORKSHEET

RADIATION: The transfer of energy as waves through space.

Describe what you felt when your hand was beside the flame.

____

How far was your hand from the flame when you could no longer feel the heat?

____

The heat that you felt from the candle was being transferred by ***radiation***. A burning candle emits electromagnetic waves that you could feel when your hand got close enough to the side of the candle. As you moved your hand away from the flame, the air absorbed the radiant heat so you could no longer feel it. If you walked 100 yards from the flame, you would still be able to see it because the air does not absorb the light energy. On a foggy day, you would not be able to see the flame from 100 yards away because the water droplets in the air reflect and absorb the light energy.

This is the same process that makes your face feel warm on a cold, sunny day. Most of the electromagnetic radiation from the sun is in the form of visible light. Solar radiation passes through the atmosphere and is absorbed by all objects: humans and other animals, plants, rooftops and other surfaces. Solar radiation has a warming effect, particularly on darker objects like asphalt driveways because they absorb more radiation and warm faster. Lighter colored objects don't absorb the sun's radiation as quickly and reflect some of it back to space.

CONDUCTION: The transfer of energy through a substance and between substances.

What happened to the paper clip in the flame?

____

What happened when you touched it to the paper?

____

The paper clip heated up because energy was transferred directly from the molecules of the hot gases in the flame to the molecules of the paper clip. The air molecules that came into contact with the hot gases of the flame were heated by conduction.

The paper clip absorbs heat energy from the flame, which makes the molecules in the paper clip move faster, warming it enough to burn a hole in the paper. This is an example of heat transfer through conduction. Heat transferred through conduction flows through a substance from warmer to colder areas and more rapidly when there are temperature differences. The rate of heat transfer through conduction also depends on whether a substance is a good conductor. Metal is a good conductor, while air is a poor conductor of heat.

Conduction can also transfer heat through two different substances that are in contact with each other. If you hold your bare foot close to a hot sidewalk on a sunny day, you might feel a little warmth due to radiant heat transfer. But if you put your bare foot on a hot sidewalk, you will feel a lot of heat because of conduction heat transfer.

CONVECTION: The transfer of energy by movement in a gas or liquid.

About how far was your hand from the flame when you first felt the heat?

What did you feel when your hand was above the flame?

The heat above the flame reached your hand as a result of convection heat transfer. The air near the flame, heated by conduction, expands and becomes less dense. Warmer, less-dense air rises. Energy is transferred upward in a convection current with the rising warmed air, which is lighter than the cooler air.

DID YOU KNOW? Notice the sha of the flame, a result of the rising ai The flame of a lit candle in a weightl environment — the International Space Station is one example — tak the shape of a ball because convecti currents do not form in a weightless environment.

The heat above a flame is hotter because of convection heat transfer. Radiation heat transfer is the same in every direction; you could move your hand as close above the flame as you did when your hand was beside the flame, if not for convection heat transfer. Conduction heat transfer doesn't come into play when your hand is above the flame because your skin is not in contact with the flame.

Convection heat transfer occurs in fluids such as water, or air, because they move freely and form currents. Convection occurs naturally in the atmosphere on a warm, sunny day as surfaces on the earth absorb sunlight and warm up. But surfaces heat unevenly. As the warmest air expands, it becomes less dense than the surrounding cooler air and rises. Rising air forms thermals, which transfer heat into the atmosphere. Cooler, heavier air then flows toward the earth's surface to replace the warm air that just rose. When the cooler air reaches the surface, it is warmed and eventually rises as a thermal, too. This circulation is called a convective circulation or thermal cell. Thermals can result in cloud formation.

What causes the shape of a flame?

MATCH THE TYPE OF HEAT WITH THE CORRESPONDING DEFINITION:

A. Radiation ______ transfer of energy by movement in a gas or liquid

B. Conduction ______ transfer of energy that requires contact

C. Convection ______ transfer of energy as waves through space

ENERGY FROM THE SUN: Energy from the sun comes in the form of radiation that can travel through the vacuum of space. Most of this radiation is in the form of visible light. Three things can happen to radiation as it strikes a substance: it can pass through, be reflected, or be absorbed.

Radiation that passes through a substance is similar to light passing through the air or a window. It is somewhat like sound waves that pass through walls so you can hear a loud radio or TV throughout your house.

Radiation can be reflected. This happens in much the same way that metal or snow reflects sunlight. Radiation can also be absorbed, particularly by dark objects. This is why a driveway heats up on a sunny day. It is also the basis for energy-efficient building materials that absorb the sun's warmth during daylight.

Some objects react with the sun's energy in more than one way. A frosted piece of glass, for example, reflects, absorbs, and allows some radiation to pass through. The earth has a variety of surfaces with diverse properties that interact with sunlight differently.

As radiation strikes a substance, what are the three possible results?

____

____

Which do you think will absorb more sunlight: snow or concrete?

____

ALBEDO: As sunlight strikes the earth, the energy is either reflected or absorbed. The ability of a substance to reflect sunlight is called ***albedo***. Some surfaces such as thick clouds and snow have a high albedo and reflect most of the sunlight back into space. Reflected sunlight has no warming effect on the air in the atmosphere.

Surfaces with a low albedo such as land and oceans absorb most of the sunlight and transform the sun's energy into heat. Absorbed heat energy warms the earth and creates the weather and climates that make life on earth possible.

Which will reflect more sunlight: a substance with a low albedo or a high albedo?

____

What happens to reflected sunlight?

____

ENERGY IN THE ATMOSPHERE: The atmosphere is composed almost entirely of nitrogen (78%) and oxygen (21)%). The amount of water vapor in the air varies, depending on conditions, but on average, water vapor is about 1.5% of the atmosphere. Carbon dioxide and other gases make up less than 0.5% of the atmospheric gases.

Gases are not affected by radiation. Sunlight passes right through gases without being absorbed. When air comes into contact with warm surfaces (such as an asphalt driveway), the surface conducts energy to the air molecules. As the air molecules get warmer, the air becomes less dense and rises. The rising air transfers energy to the upper atmosphere through conduction and convection, via thermals.

However, greenhouse gases in the atmosphere absorb infrared radiation.

Nitrogen and oxygen are very stable molecules and are not affected by radiation. They can only be warmed by conduction. Greenhouse gases like carbon dioxide are more loosely organized and are able to absorb infrared radiation. Other greenhouse gases are water vapor, methane, and nitrous oxide. All of these gases can capture the heat from the surface and re-emit that energy into the atmosphere and back toward the surface.

Earth would be a much cooler place without greenhouse gases that retain heat. The fact that they are only a small percentage of the atmosphere illustrates the power of the greenhouse effect.

How does sunlight affect the molecules in the atmosphere?

____

How does infrared radiation affect oxygen and nitrogen molecules?

____

What What percentage of the atmosphere is greenhouse gases??

____

What might a rise in greenhouse gases do to the temperature of the atmosphere?

____

EARTH'S ENERGY BALANCE

How does the sun's energy move in and out of earth's atmosphere?

RECOMMENDATION: Before you begin this activity, you may wish to review the Level 2 Weather and Climate Science activity, The Greenhouse Effect, and the Level 3 activity, Energy in the Atmosphere.

Life on earth requires energy from the sun. Radiation from the sun provides the energy that determines earth's weather and climate. Surfaces on the earth are warmed by the sun's radiation. Sunshine is essential for the health and growth of plants and animals. Some solar radiation (shortwave radiation) is transformed to heat (longwave radiation) and reflected back into space. We call the movement of the sun's energy into and out of earth's atmosphere the energy balance.

To maintain stable temperatures and life on earth, the energy entering its atmosphere must equal the energy that leaves.

OBJECTIVE: Describe the movement of energy in and out of the atmosphere.

Weather Tote

- The Energy Balance worksheet on pages 180–185

- Complete the worksheet.
- Discuss the Chat questions with a parent or other adult helper.

SHARE WHAT HAPPENED: Describe the earth's energy balance.

APPLY: How does the energy balance help explain climate change?

GENERALIZE TO YOUR LIFE: Why is it important for people to understand things that affect the earth's energy balance?

Fly Higher

Read more about the energy flux:

- NOAA:

  Earth-Atmosphere Energy Balance: www.noaa.gov/jetstream/atmosphere/energy

  The Earth's Energy Budget, https://gml.noaa.gov/outreach/info_activities/pdfs/CTA_how_much_is_the_global_energy_budget.pdf
- NASA, Climate and Earth's Energy Budget, http://earthobservatory.nasa.gov/Features/EnergyBalance/
- Climate Change and Religion, https://billmoyers.com/episode/climate-change-faith-and-fact/

EARTH'S ENERGY BALANCE WORKSHEET

The earth's energy balance describes the flow of energy into earth and back out into space. Energy from the sun flows to earth as shortwave radiation and back into space as infrared longwave radiation. The incoming and outgoing energy must be equal for the earth's temperature to remain stable over long periods of time. The state of balance is called radiative equilibrium, radiation balance, or just energy balance. If energy flow is not balanced, the earth will eventually heat up or cool down. So changes in earth's energy balance cause changes in our climate.

The earth receives about 340 W/m2 (watts per meter squared) solar radiation. Almost a third (29%) of the solar energy that arrives at the top of the atmosphere is reflected back to space by clouds, atmospheric particles, or bright ground surfaces like sea ice and snow. This energy plays no role in earth's climate system. Nearly a quarter (22.5%) of incoming solar energy is absorbed in the atmosphere by water vapor, dust, and ozone. Almost half (48%) passes through the atmosphere and is absorbed by the surfaces on the earth.

What percentage of the incoming solar energy becomes part of earth's energy system?

____

Earth's Energy System

Earth's atmosphere and surfaces absorb incoming solar radiation at different rates. Satellites are used to measure the long-term averages shown below.

	SUNLIGHT ABSORBED		HEAT EQUIVALENT RADIATED	
Measurement	%	W/m2	%	W/m2
Atmosphere	23	77	59	200
Surfaces	48	163	12	40
Total	71	240	71	240

What has the most solar heating, the atmosphere or earth's surfaces?

- Atmosphere
- Surfaces

What has the most radiative cooling, the atmosphere or earth's surfaces?

- Atmosphere
- Surfaces

How much radiation (heat equivalent percentage) must be radiated back into space for earth's temperature to remain stable over long periods of time?

____

Create two pie charts to show the amount of sunlight absorbed and heat equivalent radiated from the atmosphere and surfaces of the earth. Include titles and units (% or W/m2).

Solar Radiation

Incoming solar radiation that reaches surfaces on the earth may be absorbed or reflected. Albedo is a measure of the reflectivity of a surface. Earth's albedo effect is a measure of how much solar radiation is reflected back into space. Albedo is given as a percentage. It is measured on a scale from zero for a perfectly black surface that has no reflection to 100% for a white surface with perfect reflection.

Albedos of typical surfaces on earth range from up to 90% for fresh snow to about 4% for charcoal, one of the darkest substances. The ocean surface and most forests have low albedos, around 5%. Earth's land areas generally have albedos between 10% and 40%. Dark soil reflects about 10% of sunlight radiation. Deserts have some of the highest albedos among landforms. The average albedo of the earth is about 30%.

Absorbed solar radiation warms the surface. These surfaces can then warm the air through conduction. Air molecules must be in contact with the surface for the energy to transfer. As the air around the surface is warmed, that air becomes less dense, rises, and transfers the energy into the atmosphere as thermals, or convection.

Sunlight also causes water to evaporate and plants to transpire. The combined effect of evaporation and transpiration is known as evapotranspiration. When water changes into a gas, latent heat is produced, and the heat energy is transformed into the kinetic energy of water vapor. When the water vapor condenses, the latent heat energy is released into the atmosphere, causing the air to warm.

Draw a bar graph that shows the albedos of the following:

- Fresh snow, dark soil, charcoal, landforms (high value), ocean surface, earth (general), landforms (low value), and forests.
- Show the albedos from highest to lowest.

Describe two results of solar radiation that is absorbed on earth's surfaces?

____

____

Heat Distribution

The heat equivalent of energy radiated from earth to space is in the form of longwave radiation, usually defined as outgoing infrared energy. Longwave radiation can be reflected directly to space, partially absorbed by the atmosphere, or reflected back to earth by clouds. Heat energy is transported between the planet's surface layers (land and ocean) to the atmosphere. Many processes distribute it, including evaporation of surface water, convection, conduction, rainfall, winds, and ocean circulation. The rate and amount of radiation absorption and heat radiated varies with location and daily, seasonal, and annual changes in the earth. These processes are affected by the albedo (reflectivity) of surfaces, cloud cover, vegetation, and land use patterns.

Areas near the equator are much warmer than areas near the North and South poles. Energy moves through the atmosphere and the oceans to equalize these differences. This movement of energy in the atmosphere is called wind and weather. When this energy intensifies in a region, storms can result. Storms contain a great amount of energy, which is sometimes released with such serious consequences as hurricanes and tornadoes.

Disturbances of Earth's Radiative Equilibrium

The earth's atmosphere is a mix of nitrogen (78%) and oxygen (21%) and trace atmospheric gases (1%). Trace atmospheric gases are often called greenhouse gases because they interact with thermal infrared energy similarly to sunshine in a greenhouse.

Greenhouse gas sources include both natural and synthetic (human-created) gases. The most common natural gases are carbon dioxide, water vapor, methane, and nitrous oxide. The most common synthetic gases are chlorofluorocarbons (CFCs), hydrofluorocarbons (HFCs), perfluorocarbons (PFCs) and sulfur hexafluoride (SF6). Atmospheric concentrations of both the natural and man-made gases have been rising over the last few centuries.

Outgoing longwave heat equivalent radiation does not fully pass through greenhouse gases in the same way it passes through nitrogen and oxygen. Nitrogen and oxygen allow outgoing radiation to pass through unchanged. Greenhouse gases, however, absorb some thermal infrared energy. These gas molecules radiate infrared energy in all directions, so some of it spreads downward and ultimately comes back into contact with the earth's surface, where it is absorbed. This causes the temperature of the earth's surface to be warmer than if it were heated only by direct solar heating. This supplemental heating of the earth's surface by the atmosphere is the natural greenhouse effect.

The earth's climate has evolved the delicate balance between incoming solar radiation and outgoing thermal radiation that supports life on earth over millennia. This balance is so delicate that changes in the greenhouse gases, only 1% of our atmosphere, can tip this balance!

Human Influences

Humans tend to change the natural environment for their use. Some changes can affect the surfaces of earth and the composition of the atmosphere, and therefore, the energy balance. Many people are advocating for more care of our natural resources to reduce human impacts on the environment.

How might the following activities affect the earth's energy balance?

Cutting down forests

____

____

Using energy sources like wind, water, and solar energy

____

____

Producing more efficient cars and trucks

____

____

Burning more fossil fuels

____

____

Constructing more buildings, roads, and parking lots

____

____

Planting more trees and creating more green space

____

____

Recycling

____

____

What other activities or concerns have you heard about or researched that might affect earth's climate?

NOTES

IMPACT OF CLIMATE CHANGE

How does climate change impact our environment?

intro Polar bears have become the environmental "poster child" of climate change. Melting Arctic ice is reducing their mobility and access to transit and hunting grounds for fish and seals. People who worry about climate change focus on the loss of polar bear habitat because polar bears are iconic to the Arctic.

Many other species will be impacted by climate change, of course. For over 100 years, scientists have realized that an increase in greenhouse gases in the atmosphere causes an increase in the average global temperature. Natural fluctuations in the climate system will continue. Winters will still be cold and some will be more severe than others. Global temperatures will not increase every year, but over time baseline temperatures will continue to climb as long as the amount of greenhouse gases increases. While scientists cannot predict exactly when the Arctic will become ice-free and the polar bear habitat will be gone, they know that it will happen if greenhouse gases emissions are not controlled. In this activity you will conduct research to find out more about the expected environmental impact of climate change on a specific animal or plant.

OBJECTIVE: Learn about the impact of climate change on an animal or plant.

Weather Tote

- Internet access

  Use multiple sources.

  Use websites ending in .edu or .gov for scientific organizations linked to university or governmental scholars.

- Materials to present your research results via print, electronic or video

  Poster presentation

  PowerPoint presentation

  YouTube video

DO it

1 Choose a plant or animal to research (your subject).

You may select any species of your choice.

Or, choose one of the following:

- Blue crab
- Coastal wetlands
- Coral reefs
- Eastern hemlock Forests
- Lake sturgeon
- Mangroves
- Moose
- Penguins
- Pika
- Polar bears
- Ringed seals
- Sagebrush
- Salmon
- Walleye
- Western forests (mountain pine beetle)

2 Take notes while you learn all you can about your subject.

- Index cards can be used to keep your notes. Put different topics on each card (for example, food and shelter needs, survival temperature range, special reproduction requirements, range) for easy sorting later.

3 Prepare a short manuscript (3–5 pages) to describe what you learned about the expected environmental impact on the subject you chose.

- Create an outline before you begin.
- Arrange the information you found during your research, following your outline.
- Write your paper (manuscript).
- Include possible solutions and/or programs to address the problem.
- Include pictures, graphs, and/or tables, as appropriate, to help your reader understand the impact of climate change on the species you researched.
- List the references you used:

  Book references must include the title, authors, publisher, and date of publication.

  Internet references must include the page title, professional affiliation (university, governmental department, or professional society), web address, and date of access.

  Articles in professional papers, magazines, or newspapers must include the title, author, source, and date of publication.

4 Prepare a poster exhibit to convey the plight of your subject to the public.

- A good poster:

  Attracts attention

  Is simple and clear

  Interests people in your subject

  Does not contain too much information for a casual reader

Chat SHARE WHAT HAPPENED:

- Why did you choose the particular species for a research subject?
- How many sources did you use?
- How long did you spend on your research?
- How long did it take you to create your presentation?

APPLY:

- How does what you learned apply to other species?
- Did you discover any actions you could take to help with the situation?

GENERALIZE TO YOUR LIFE: What effect might climate change have on your life?

Fly Higher

- Study the impact of climate change on another species.
- Visit the United Nations Climate Change webpage, http://www.un.org/climatechange/, to study what world leaders are thinking and doing about the problem.
- Visit the Purdue Climate Change Research Center (PRCC), https://www.adaptationclearinghouse.org/organizations/purdue-climate-change-research-center-pccrc.html, to learn about the research it is doing.
- Investigate other issues related to environmental impact of climate change.

INVESTIGATING CLIMATE CHANGE

How can we use climate data to demonstrate cause-and-effect relationships?

intro Many tools can be used to observe and track the weather. Thermometers, rain gauges, anemometers, and wind vanes are used to determine the local weather on a day-to-day basis. Weather reports and maps can be used to monitor weather on a larger scale. But how can *climate* be monitored?

Climate must be observed over a period of years. Weather records are used to develop climate averages. Past data can be observed and used to predict future weather patterns. Climate norms are based on 30-year data sets. The National Centers for Environmental Centers generates the official U.S. normals every 10 years (https://www.ncei.noaa.gov/products/land-based-station/us-climate-normals). Weather data is tracked and reported monthly, yearly, or every few years depending on the type of data. Much of this data has been tracked for over 100 years.

Many organizations report weather data. The National Oceanic and Atmospheric Administration (NOAA), the Environmental Protection Agency (EPA), and the National Aeronautics and Space Administration (NASA) are reliable sources for weather data. While many websites may change over the years, the climate information from the NOAA, EPA, and NASA websites has remained fairly consistent.

OBJECTIVE: Use a government weather data source to record the weather climate data, and describe the cause-and-effect relationships that you observe.

Weather Tote

- Materials to create graphs and charts, such as paper and pencils or a Microsoft Excel spreadsheet
- Materials to create a display, such as the following:
  - Markers
  - Graphics or pictures
  - Computer-generated titles
  - Poster board/trifold board

Do it

1. Visit the NOAA, EPA, and NASA websites shown in Table 16.
   - Investigate the data available at each website.
2. Decide on two weather measurements that you will list for each year over a 30-year period (ending in the current year).
   - Choose data that you feel is interesting, relative, and significant.
   - Choose data that can show cause-and-effect relationships.

3 Decide on a location such as your home or a major city.

4 Create a table that lists the data for each of the last 30 years for the two measurements.

5 Graph your data using a line graph or bar graph. You can do this by hand or use Excel.

6 Discuss any trends that your data shows over the 30-year time span.

7 Prepare an educational poster, showing:

- The data you collected (include units!)
- Your discussion of current trends in the data
- Your explanation of how the data sets compare and/or relate to other data sets that people may recognize
- A good poster:

  Attracts attention

  Is simple and clear

  Interests people in your subject

  Does not contain too much information for a casual reader

8 Present your poster at one of the following venues:

- Demonstrations (including 4-H interactive demonstrations)
- Information display boards
- Junior Leader activities
- Mentoring a younger 4-H member
- Outreach to community members

SHARE WHAT HAPPENED: Where did you collect your weather data? Why?

APPLY:

- How does knowing weather trends apply to your daily life?
- How do you think your results would look if you were a 4-H member in 1970 and collected the previous 30 years of weather data?
- How will our idea of climate change differ in the next 30 years?

GENERALIZE TO YOUR LIFE: Why is it important that others understand weather trends?

Fly Higher

- Continue tracking your data throughout your 4-H membership and beyond.
- Choose other weather topics to track using historical data.

TABLE 16. **INVESTIGATING CLIMATE CHANGE WEBSITES**

Provider	Website	Website
NOAA	www.ncdc.noaa.gov/	National Climate Data Center
EPA	https://19january2021snapshot.epa.gov/climate-indicators/frequent-questions-about-climate-change-indicators_.html	Climate Change Indicators in the United States (temperature)
NASA	www.earthdata.nasa.gov/topics/climate-indicators	Global Change Master Directory

THE SUN–EARTH RELATIONSHIP

What caused the ice ages?

intro The earth's temperature has varied significantly during the millennia, from the ice ages to a period when tropical temperatures blanketed the entire earth. Scientists have tried to explain this variation by studying the relationship between the sun and earth. Milankovitch theorized that the direction and angle of the earth's tilt, and the shape of the orbit are responsible for the significant temperature changes on a millennia scale. In this activity, you will investigate the Milankovitch Cycles.

OBJECTIVE: How do Milankovitch Cycles describe the earth's orbit and orientation over time?

Weather Tote

- Milankovitch Cycles—The Sun-Earth Relationship worksheet on pages 191–197

Do it

1. Complete the worksheet.
2. Discuss the Chat questions with a parent or other adult helper.

Chat

SHARE WHAT HAPPENED: What are the Milankovitch Cycles?

APPLY: Why is it warmer in the summer than in the winter in the United States?

GENERALIZE TO YOUR LIFE: Events that follow cycles will occur again. How can genealogy describe a family cycle?

THE SUN–EARTH RELATIONSHIP WORKSHEET

Earth's temperature has varied significantly since it was formed. At least 30 glacial periods have occurred. Figure 68 shows temperature and carbon dioxide (CO_2) data from ice core samples that chronicle the temperature and carbon dioxide levels during the past 400,000 years. There is a direct correlation between the earth's average temperature and the occurrence of ice ages. The lower blue line on the graph indicates the temperature deviation in °C. The upper red line shows CO_2 levels. The CO_2 level was at 380 ppm when this graph was created in 2007. In 2013 the CO_2 level was 399 ppm. The average global temperature has varied about 8°C during the past 20,000 years.

FIGURE 68. **VARIATIONS IN TEMPERATURE AND CARBON DIOXIDE**

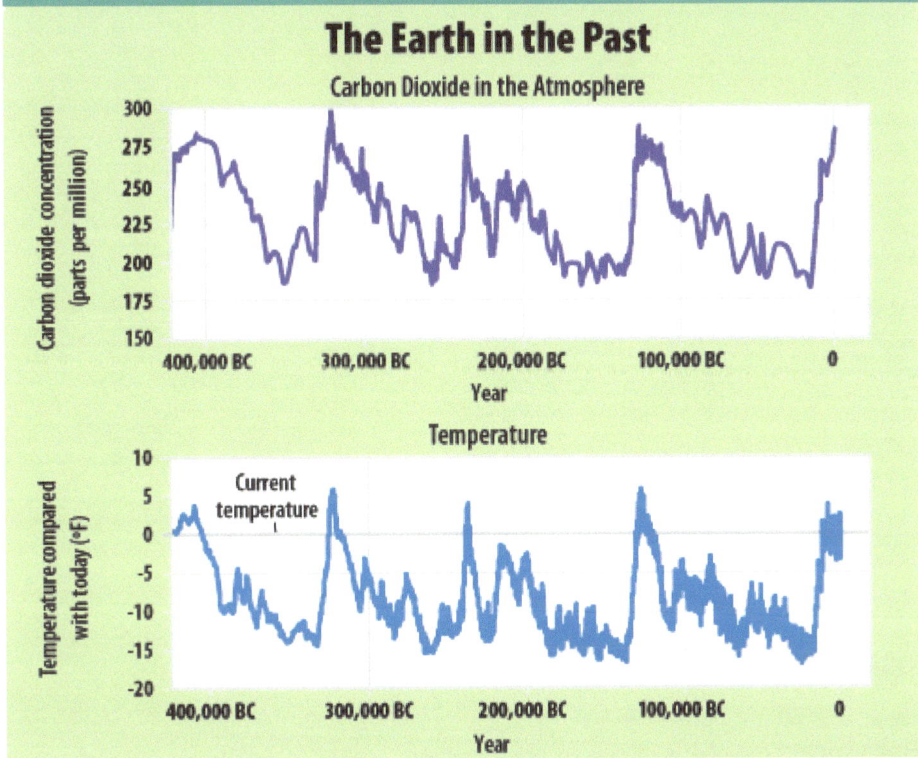

Source: U.S. EPA

How many cooling periods do you observe during the past 400,000 years in Figure 68?

____

Which change seems to occur at a faster rate, warming or cooling?

____

Milankovitch's Theory

A Serbian scientist named Milutin Milankovitch developed a theory in the 1920s to explain how earth's changing patterns of orbit and tilt in relationship to the sun might cause climate change. He was especially interested in explaining the occurrence of the ice ages. Milankovitch suggested that changes in the earth's orbit and tilt occur in cycles over very long periods of time—tens of thousands of years. These cycles, based on mathematical calculations, are now known as the Milankovitch Cycles. The scientific community did not take his theory seriously until the 1970s, when actual ice core data showed a possible connection between the ice ages and the Milankovitch Cycles.

Milankovitch studied three aspects of the earth's orbit around the sun that influence the amount of energy the earth receives from the sun. The cycles are obliquity, eccentricity, and precession. Could these cycles help explain the recent rise in the earth's temperatures?

What did Milankovitch's theory try to explain?

____

OBLIQUITY

Seasonal changes are caused by the tilt of the earth. The earth's tilt is currently about 23.5° when compared with its orbit around the sun. The first day of summer in the Northern Hemisphere occurs when the North Pole is tilted at the maximum 23.5° toward the sun. On that day, the Northern Hemisphere receives the maximum amount of the sun's radiant energy. The day also marks the first day of winter for the Southern Hemisphere. The angle created by the earth's tilt varies slowly between 22.1° and 24.5° over a time span of approximately 41,000 years. The earth's tilt is responsible for the changing seasons and their differences in temperature.

ECCENTRICITY

The earth orbits the sun following an elliptical (oval) path. Eccentricity is a measure of the departure of the earth's orbit from a circle. The shape of the earth's orbit varies at a slow rate, between nearly circular (low eccentricity of 0.005) and mildly elliptical (high eccentricity of 0.058) with a mean eccentricity of 0.028. The elliptical orbit causes the distance of the earth from the sun to continuously change throughout the year. The earth's average distance from the sun is 93 million miles. At earth's closest, it's about 91.5 million miles from the sun. This occurs annually in early January.

The earth reaches its farthest position from the sun in early July, when it is about 94.5 million miles from the sun. Some evidence exists that this elliptical orbit may make the Northern Hemisphere seasons less severe than seasons in the Southern Hemisphere, although the difference is thought to be slight.

PRECESSION

Precession is a measure of the direction of the earth's axis of rotation relative to the fixed stars. Currently, the North Pole is pointing toward the North Star. This causes the Northern Hemisphere to tilt away from the sun when the earth is closest to the sun (our winter) and toward the sun when the earth is farthest from it (our summer). But this is not a constant. Precession changes the tilt very slowly, over a period of about 26,000 years. The North Pole changes its orientation as the earth spins, so it is not always pointed toward the North Star.

Which cycle explains the direction of the earth's tilt?

____

Which cycle explains the shape of the earth's orbit?

____

Which cycle explains the angle of the earth's tilt?

____

How might the Milankovitch Cycles influence present-day global temperatures?

ECCENTRICITY

Eccentricity is a measure of how much the ellipse of the earth's orbit differs from a circle. The ellipse on the left in Figure 69 nearly circular, while the one on the right is more oval. Therefore, the ellipse on the right is more eccentric than the ellipse on the left.

FIGURE 69. **CIRCLE AND OVAL**

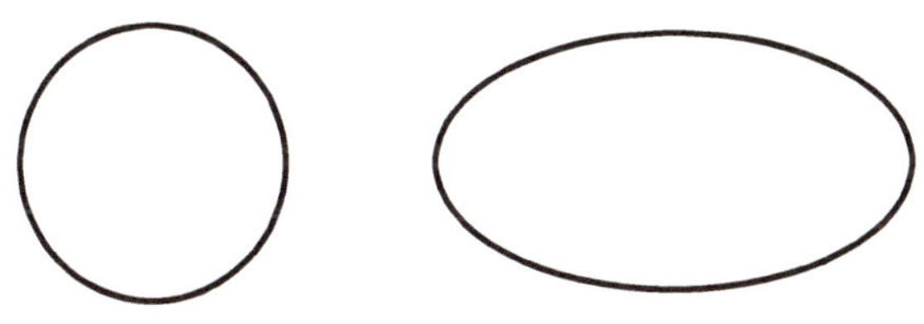

Because its orbit is an ellipse, there are times when the earth travels closer to the sun. Annually, the earth's elliptical orbit varies from a circular shape by three million miles. The average radius of the earth's path is 93 million miles, so the variation from a circular pattern is slight. This amount of change is relatively small and does not seem to cause significant temperature changes on earth.

Milankovitch discovered that the shape of the ellipse changes and becomes more or less eccentric over time. A more circular orbit would create less severe seasons and therefore less seasonal change. A less circular orbit would cause the earth to be much farther away from the sun at the farthest point of the orbit than it is now. A large change could cause more severe seasons. Cooler summers could cause the buildup of snow and ice fields similar to the glacial periods of the ice age. The eccentricity cycle is estimated to repeat every 100,000 years.

What is the percentage of eccentricity of the earth's orbit?

___

What effect would a more circular orbit have on seasons?

___

What effect would a more eccentric orbit have on seasons?

___

How long does it take for this cycle to repeat?

___

OBLIQUITY

Obliquity is the tilt of the earth in relationship to the sun. It changes over time. At present the tilt of the earth is 23.5° in respect to its orbit around the sun. The earth's tilt is responsible for the changing seasons and their differences in temperature. Through a 40,000-year period, the obliquity (tilt) of the earth changes from a minimum of 22.1° to a maximum of 24.5° and back again. This change in tilt occurs very slowly, but could cause the seasons to be more severe during maximum tilt and less severe at minimum tilt.

When the earth is only tilted 22.1°, summers and winters may be less severe. At the minimum tilt, higher latitude summers would be cooler and winters warmer. Scientists speculate that, during such cooler summers, winter snow and ice in the higher latitudes might not melt from year to year. In addition, the buildup of snow and ice fields could cause more of the sun's energy to be reflected back into space, promote the formation of glaciers, and possibly result in an ice age.

At maximum tilt, seasonal changes would be more dramatic. Colder winters and warmer summers for the higher latitudes would be likely. Studying Figure 70 may help you understand this relationship.

FIGURE 70. EARTH'S ORBIT AROUND THE SUN

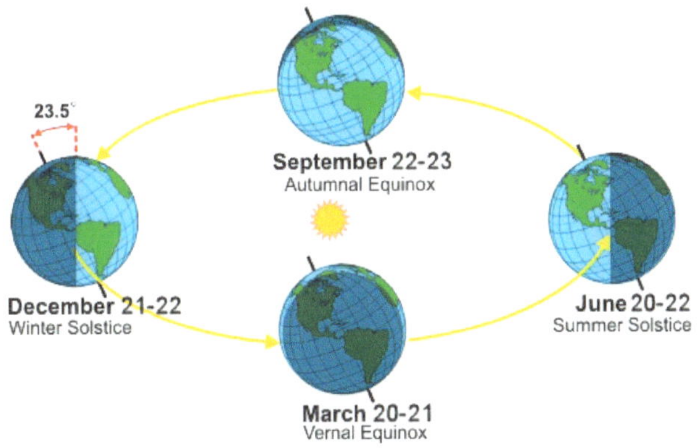

At a maximum tilt of 24.5°, what will happen to seasons in the higher latitudes?

____

At a minimum tilt of 22.1°, what will happen to seasons in the higher latitudes?

____

How long does it take for earth to complete the tilt cycle?

____

PRECESSION

Currently the earth's axis is oriented so that it points almost directly at the North Star, Polaris. Four thousand years ago, when Egypt was building the pyramids, the North Pole was oriented to the star Thuban in the constellation Draco. The earth's orientation has shifted over time. This shift is called precession. The North Pole will shift following a 26,000-year cycle. So 26,000 years from now, the North Pole will once again be aligned with Polaris.

Today's precession orientation causes the earth to be nearest the sun during our winter and farthest away during our summer. This causes less severe seasons in the Northern Hemisphere.

In about 12,000 years, the precession cycle will move the tilt so the earth will be nearest to the sun during our summer and farthest away during our winter. This precession position could cause more severe seasons in the Northern Hemisphere.

During which of the Northern Hemisphere seasons is the earth closest to the sun?

How long does this cycle take to complete?

When the earth is closest to the sun during our summer, what effect might that have on our seasons?

How do the Milankovitch Cycles combine to affect climate?

Scientists speculate that the most likely scenario for a Northern Hemisphere ice age would occur under the following combination of Milankovitch factors:

- A nearly circular shape of the earth's orbit (minimum eccentricity)
- A minimum tilt of 22.1° (minimum obliquity)
- Orientation so earth is closest (and farthest) away from the sun during spring (and autumn) (precession)

According to Milankovitch's theory, these factors together would cause the four seasons to be similar to each other. The resulting seasonal temperature changes, even though very small, could allow areas near the poles and higher latitudes to accumulate snow and ice and to form glaciers. Increased snow and ice on the earth would reflect more sunlight and the earth would cool. The opposite cyclical extremes could cause a warm period. This warming would cause the snow and ice fields to shrink and the earth to absorb more of the sun's energy.

Scientists generally think that these cycles will combine to produce either a cooling trend or a warming trend. Currently scientists are predicting that the earth's temperatures should remain stable before moving toward an ice age, which could occur in about 30,000 years. Human impacts, such as an increase in CO_2, may affect the expected pattern.

What status of the Milankovitch Cycles are expected to create the next ice age?

Could these cycles cause changes during a period of hundreds or thousands of years? Why?

FIGURE 71. DEVIATION FROM ANNUAL MEAN TEMPERATURE

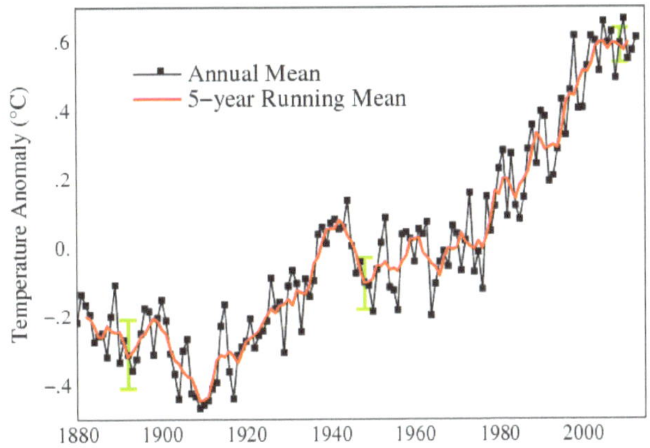

Figure 71 shows a rise in global temperatures compared to the global average since 1880. Could changes in the Milankovitch Cycles explain the recent rapid global temperature rise shown in Figure 71? Why or why not?

____

____

Figure 68 shows the temperature has risen about 8°C over the past 20,000 years. Figure 71 shows the temperature has risen over 1.5°C in the last 130 years. Which time period is showing the most rapid average temperature rate?

____

Climates and seasons on earth are primarily a function of how much solar radiation is absorbed and retained by the different areas. Many factors help determine exactly how solar radiation affects the climate. There is a natural relationship between the earth and the sun in space. Seasons are determined by the position of the earth in respect to its orbit and tilt. The distance of the earth from the sun is also important. When the earth is closer, the increased radiant energy can affect our climate. These factors also can change over time within recurring cycles.

A correlation was shown between the Milankovitch Cycles and the occurrence of ice ages. The Milankovitch Cycles of eccentricity, obliquity, and precession predict that ice ages should occur when all three encourage less severe seasons.

At present, only one cycle, precession, is at an extreme. The current precession of the earth places it closest to the sun during our winter and farthest away during our summer. This should make the Northern Hemisphere's seasons milder, but conversely, the Southern Hemisphere's seasons should become more severe. The obliquity, or tilt, of the earth is currently at neither extreme. The 23.5° of tilt is midway between the extremes of 22.1° and 24.5°. The eccentricity of the earth's orbit also is not at an extreme. The Milankovitch Cycles do not explain the current increase in global temperature.

SUNSPOT CYCLE

What is the sunspot cycle?

Global warming and climate change are international concerns and the focus of much controversy. In this activity, you will learn about the **sunspot cycle**, which causes increases and decreases in the radiation from the sun to the earth. You will use this information to reflect on the relationship between sunspots and global temperatures.

OBJECTIVE: Relate the relationship of the sunspot cycle to global temperatures.

Weather Tote

- Sunspot Cycle worksheet on pages 199–204

1. Complete the worksheet.
2. Discuss the Chat questions with a parent or other adult helper.

SHARE WHAT HAPPENED: What are sunspots?

APPLY: What happens to the earth when there is very little sunspot activity?

GENERALIZE TO YOUR LIFE: Are an increased number of sunspots (solar max) responsible for climate change?

Fly Higher

View the current state of the sun and recent solar activity captured by the Solar and Heliospheric Observatory (see http://sohowww.nascom.nasa.gov/sunspots/) for one week. Describe your observations.

SUNSPOT CYCLE WORKSHEET

Global warming and climate change are international concerns and the focus of much controversy. Strong scientific evidence shows an accelerated rise in average global temperatures during the past 30 years. The media gives a great deal of attention to the fact that concentrations of greenhouse gas and carbon dioxide (CO_2) are rising.

The central cause of this increase is usually blamed on human activity, especially the burning of fossil fuels, which puts large amounts of CO_2 into the earth's atmosphere. The increase in carbon dioxide is the most frequently discussed factor in the global-warming controversy.

Many variables actually contribute to the temperature changes associated with global warming, and they all must be part of a thorough investigation of it.

One of these variables is **sunspot** activity. Energy from the sun produces the heating that occurs on the earth. The energy coming from the sun is not always the same. Sunspots on the surface of the sun indicate an active state (see Figure 72). When there are few or no sunspots present, the sun is in a less active state and emits less energy. The **sunspot cycle** causes increases and decreases in the radiation from the sun to the earth.

FIGURE 72. **SUNSPOT**

Think About It—Your Best Guess.

How do you think sunspot activity affects the amount of solar radiation given off by the sun?

____

How might an increase in radiation from the sun affect the earth's temperature?

____

How might the sunspot cycle explain recent increases in global temperature?

____

FIGURE 73. **DEVIATION FROM ANNUAL MEAN TEMPERATURE**

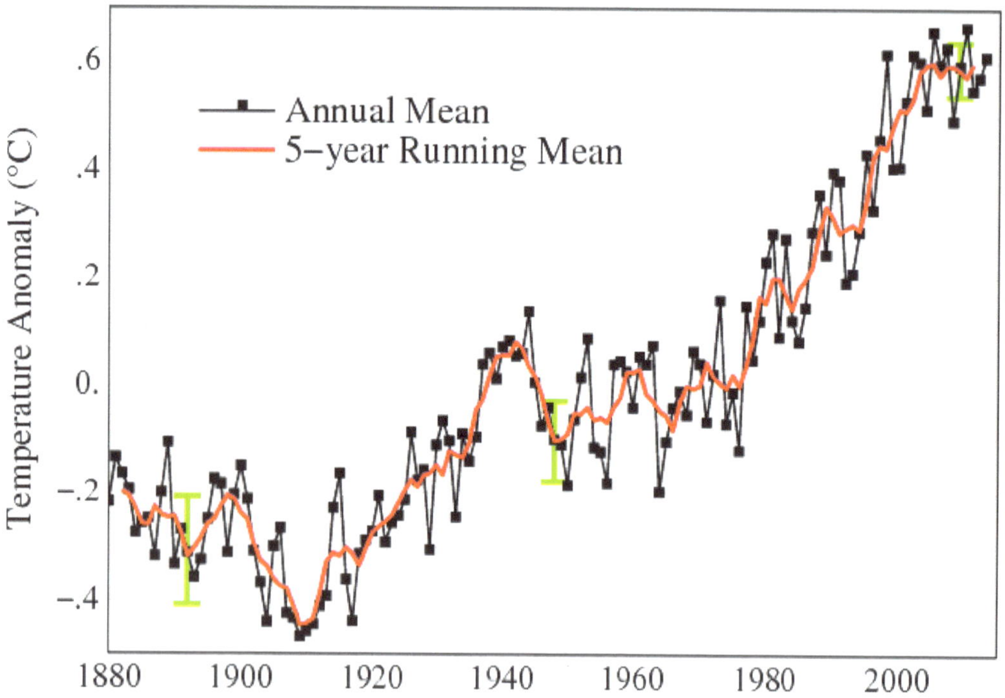

Figure 73 shows the deviation from the mean global temperature for the years from 1880 to 2007. When the yearly average equals the mean, that year is graphed at 0. When the yearly average is below the mean, the deviation is below 0; and when it is above the mean, the deviation is above 0. So Figure 68 shows how the earth's overall average temperature compares to each year's average temperature. Use Figure 73 to answer the following questions.

Were temperatures increasing, staying the same, or decreasing from 1880 through 1920?

___

Were temperatures increasing, staying the same, or decreasing from 1920 through 1977?

___

Were temperatures increasing, staying the same, or decreasing from 1977 through 2012?

___

During which time period do you see the most change occurring?

___

Which five years were the warmest?

___

About every 11 years, changes in the sun's magnetic field produce a **solar maximum** or "**solar max.**" A high number of sunspots creates a solar max, when activity on the sun is extraordinary. In addition to sunspots, there is an increase in solar flares, prominences, and corona mass ejections.

Sunspots are areas that are cooler than the areas around them. They occur when the magnetic field of the sun becomes very active. During a solar max, other areas become brighter, and solar storms are very common. This activity results in the sun's release of massive amounts of energy into space. The energy bombards the earth's atmosphere, resulting in many outcomes, some negative and some positive.

On the negative side, cell phone and radio reception can be interrupted; satellites can be rendered useless. Positives include the wonderful "sky shows" observed as the earth's magnetic field draws in the solar wind, producing the northern lights. The most significant effect of a solar max is the extra amount of radiant energy the sun sends to the earth.

In contrast to the solar maximum, a **solar minimum** occurs when few or no sunspots are found on the sun. During a solar minimum, the sun emits less radiant energy.

What effect do you think a solar maximum has on temperatures on earth?

____

What effect do you think a solar minimum has on temperatures on earth?

____

Galileo observed sunspots in 1610 and began what would become an almost complete record of sunspot activity. It was not understood at that time that sunspot activity was related to solar maximums (high sunspot activity) and minimums (low sunspot activity).

The years between 1645 and 1715 are referred to as the Little Ice Age. The earth was unusually cold during this period. Normally ice-free rivers froze, and snow fields lasted year-round in areas where snow normally melted in the summer. The period corresponded to a time of very little sunspot activity. Figure 74 shows the yearly average number of sunspots between 1610 and 2000.

FIGURE 74. **YEARLY AVERAGED SUNSPOT NUMBERS, 1610–2010**

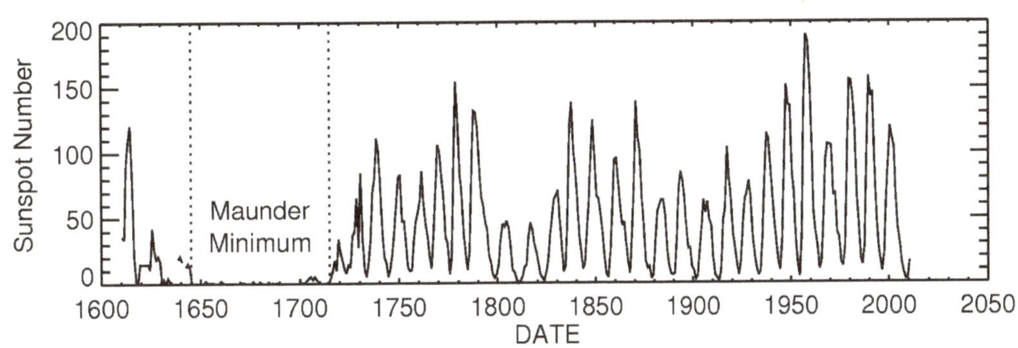

Source: NASA

How might sunspot activity have contributed to the Little Ice Age?

Sunspot activity from 2000 through 2014 is shown in Figure 75. The most recent cycle peaked in the summer of 2013. From March 2006 through the end of 2009, the sun went through a period of sunspot inactivity. Figure 76 shows the average global temperature from 1998 through 2010. Compare the recent sunspot activity to the global temperatures for 2000–2014 and answer the following questions.

FIGURE 75. **SUNSPOT CYCLES, 2000–2014**

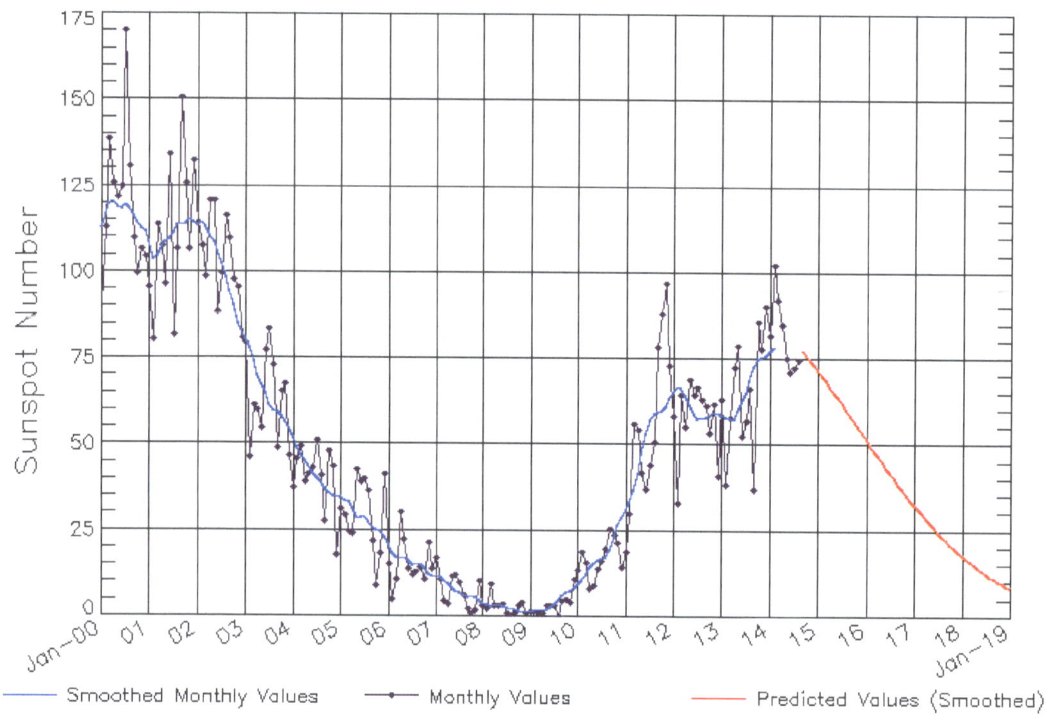

Source: National Oceanic and Atmospheric Administration/Department of Commerce.

Was sunspot activity increasing or decreasing from 2000 to 2014?

Was global temperature increasing or decreasing from 2000 to 2010?

What should have happened to global temperatures between 2000 and 2010 as a result of sunspot activity?

FIGURE 76. **MEAN ANNUAL TEMPERATURE**

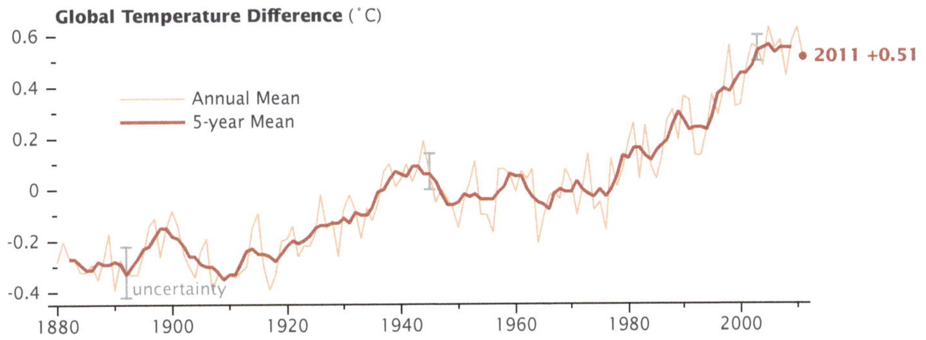

Source: NASA

The earth experienced three of the highest recorded temperatures in 2005, 2007, and 2010. What was the sunspot activity like during these years?

____

Does sunspot activity explain the most recent increase in temperature? Why or why not?

____

Climate and seasons on the earth are primarily a function of the amount of solar radiation that is absorbed and retained by the earth, but many other factors help to determine exactly how solar radiation affects climate. The amount of solar radiation given off by the sun will always directly affect the climate on earth. The sun's energy output is not steady, however, but changes over time within a cycle of maximum and minimum output.

A combination of factors contributes to our climate. These factors can occur as cycles or single events and complicate the study of climate change. The eruption of Mount Pinatubo, a volcano in the Philippines, in 1991 is an example of a single event. This eruption was the second largest volcanic eruption of the twentieth century. Within two hours of the major Pinatubo eruption, gases, and ash reached high into the atmosphere. By the end of that year, the eruption had created a high-altitude particulate cloud that covered the whole earth. These particulates reflected sunlight into space and caused a global cooling. This cooling effect lasted through 1992 and 1993 and may have caused the drop in temperatures observed during this time period. Yet questions remain. Was the drop in temperatures in 1992 and 1993 due to the volcano, or the solar minimum, or both? An El Niño event also occurred in 1993. What effect might the El Niño event have had on global temperatures? If greenhouse gases had not been involved, would temperatures have dropped even more? If there had been no volcanic eruption, would temperatures have risen during those years? Were other factors involved?

While many factors contribute to weather and, therefore, climate, scientists studying climate change focus on factors that are consistently increasing to explain the rising average global temperature. While there are variations due to other cycles and events, greenhouse gases, especially carbon dioxide (more than 30 billion tons/year), have been steadily rising so are considered the primary reason for the increasing average global temperature.

How does sunspot activity affect the amount of solar radiation given off by the sun?

____

How might increased radiation from the sun (sunspot cycle) affect the earth's temperature?

____

Can the sunspot cycle alone explain recent increases in global temperature?

____

What other factors could contribute to global temperatures?

____

If Mount Pinatubo hadn't erupted in 1991, what might have happened to temperatures?

____

What is the only persistent change that could explain the increases in temperature?

____

VOLCANOES AND GLOBAL WARMING

Do volcanoes have a significant effect on global temperatures?

Are volcanoes a source of atmospheric carbon dioxide? This question is a point of controversy related to the greenhouse effect and its potential impact on global warming. In this activity, you will discover how volcanic eruptions contribute to atmospheric carbon dioxide, the greenhouse effect, and global temperatures.

OBJECTIVE: Compare and contrast the effects of human activity and volcanic activity on the temperature of the atmosphere.

Weather Tote

- Volcano and Global Warming worksheet on pages 206–209
- Graph paper and pencil

1. Complete the worksheet.
2. Discuss the Chat questions with a parent or other adult helper.

SHARE WHAT HAPPENED: Do volcanoes cause global warming? Can they cause global cooling?

APPLY: How does an understanding of chemistry help someone understand climate change?

GENERALIZE TO YOUR LIFE: Why do some people make arguments that are not based on data?

Fly Higher

- Identify three recent volcanic eruptions. Investigate the effect they had in their own regions and worldwide.
- Look online for information about supervolcanoes.

VOLCANOES AND GLOBAL WARMING WORKSHEET

Volcanoes' contribution to carbon dioxide in the atmosphere is a point of controversy in the discussion of the greenhouse effect and its potential impact on global warming. In this activity, you will learn how volcanic eruptions contribute to atmospheric carbon dioxide, the greenhouse effect, and global temperatures.

Volcanoes eject many different materials into the air when they erupt, including a variety of gases such as water vapor and carbon dioxide, and ash (small particles of dust). Water vapor and carbon dioxide are considered greenhouse gases. Another important gas produced by volcanoes is sulfur dioxide, which is not a greenhouse gas but has an effect on the atmosphere.

What do you think?

Which do you think contributes more carbon dioxide per year into the atmosphere, volcanic eruptions or human activity? Support your response.

____

____

____

Do you think volcanic eruptions cause atmospheric warming or cooling?

____

Explore and Explain

Large volcanic eruptions put water vapor (H_2O), carbon dioxide (CO_2), sulfur dioxide (SO_2), and other gases and sometimes large amounts of ash into the air. Carbon dioxide and water vapor are greenhouse gases that contribute to global warming.

Water combines with other volcanic gases to form hydrochloric acid (HCl) and hydrofluoric acid (HF) that fall as acid rain. Sulfur dioxide converts to a fine mist of particulates called **aerosols.** These sulfur dioxide particulates (sulfate aerosols) migrate high into the atmosphere and reflect sunlight into space. Sulfate aerosols encourage the formation of high clouds that also reflect sunlight into space. The aerosols and clouds reduce the amount of radiation warming and cause the atmosphere to cool.

Sulfate aerosols tend to stay in the upper atmosphere for months or even years before they are brought back to earth in the form of sulfuric acid. Volcanic ash in the atmosphere blocks sunlight from reaching the surface of the earth, causing the air to cool.

Which greenhouse gases are added to the atmosphere during a volcanic eruption?

___

How do sulfate aerosols affect the temperature of the atmosphere?

___

How does volcanic ash affect the temperature of the atmosphere?

___

Looking at data:

Many factors must be considered in the study of volcanic eruptions and their impact on the atmosphere. Volcanoes are not all the same.

The type of eruption depends on the circumstances that cause the volcanic activity in the first place. Volcanoes created by hotspots in the ocean are rarely explosive and do not eject much ash. Eruptions taking place at tectonic plate boundaries, however, can be highly explosive with a great deal of ash.

The types and amounts of gases ejected by each type of volcano differ, too. Table 17 shows the gases ejected by three different volcanoes in three different geographic locations.

TABLE 17. **GASES EJECTED BY THREE VOLCANOES**

Volcano Tectonic Style Temperature	Kilauea Hotspot 1170°C	Erta Ale Divergent Plate 1130°C	Momotombo Convergent Plate 820°C	Average (%)
H_2O	37.1 %	77.2 %	97.1 %	
CO_2	48.9 %	11.3%	1.4%	
SO_2	11.8%	8.3%	0.5%	

Calculate the average for each gas and enter the value in column 5 of Table 17.

___

On average, which gas—H_2O, CO_2, or SO_2—composes the highest percentage of the ejected gases in the volcanoes listed in Table 17?

___

Which volcano ejected the greatest percentage of carbon dioxide?

____

Which volcano ejected the least amount of carbon dioxide?

____

Why might there be so much variation in the composition of gases that come from volcanic eruptions?

____

____

Some people argue that volcanic eruptions are a major cause of carbon dioxide levels in our atmosphere. USGS (United States Geologic Service, http://volcanoes.usgs.gov/hazards/gas/climate.php) carbon dioxide emission estimates show, however, that human activity outpaces volcanic activity. Consider that the annual average carbon dioxide emissions from human activity is about 30 billion metric tons (30,000,000,000 or 3.0×10^{10} tons). The annual average carbon dioxide emissions from volcanic activity is 145 – 225 million metric tons (average of 185,000,000 or 1.85×10^{8} tons).

How much more CO_2 do humans create than volcanoes (percent)?

____

How much CO_2 do volcanoes create compared to humans (percent)?

____

Compared to human activity, how important are volcanoes in adding carbon dioxide to the atmosphere?

____

Mount Pinatubo is a volcano that lies near a divergent plate boundary in the Philippines. Mount Pinatubo violently erupted in June 1991 and is now on record as the second largest volcanic eruption of the 20th century.

The eruption had 10 times the explosive power of the 1980 eruption of Mount St. Helens in the state of Washington. Within two hours of the major Pinatubo eruption, gases and ash had reached high into the atmosphere. During the following two weeks, the sulfate aerosols created by the blast encircled the globe. At the end of a year, the entire atmosphere of earth was filled with a layer of sulfate aerosols ejected from the eruption.

The sulfur aerosols created by the Pinatubo blast reflected solar energy into space and decreased temperatures around the world in 1992 and 1993. In spite of rising amounts of greenhouse gases and the presence of an El Niño event—factors that would normally warm the atmosphere—the sulfate aerosols reduced global temperatures by about 0.4 to 0.5°C (0.7–0.9°F) globally. The United States experienced its third coldest summer in 77 years in 1992.

The 1815 eruption of the Tambora Volcano in Indonesia created an even larger global cooling effect. Global temperatures were lowered by as much as 3°C (5.4°F). 1816 was known as the year without summer in many parts of Europe and North America, and Indiana experienced snowfall in June and frost in July that year.

What is the effect on global temperatures of the sulfate aerosols formed by volcanoes?

____

Using the ideas of greenhouse gases, volcanic ash, and sulfate aerosols, summarize the effect of volcanoes on global temperatures.

____

____

____

____

____

GLOSSARY

Absorb: To take in all or part of a substance: The earth's surfaces absorb radiant energy.

Aerosol: Particles suspended in the air.

Air mass: A large body of air with similar temperature and moisture characteristics.

Air pressure: The force that air exerts on any surface in contact with it.

Albedo: The ratio of reflected light energy to absorbed light energy; ability to reflect sunlight.

Albedo effect: A measure of how much solar radiation is reflected.

Atmosphere: A layer of gases surrounding the earth, held in place by gravity.

Barometric pressure: Atmospheric pressure as measured by a barometer, usually in millibars or inches of mercury.

Calibrate: To check and adjust a measurement precisely against a standard measure.

Climograph: A graphic depiction of the precipitation and temperature for a certain location.

Clockwise: The rotational direction of the hands of a clock.

Cold front: A front formed by a cold air mass pushing into an area of warmer air.

Concentration: The amount of a substance in a given amount of another substance.

Condense: To convert a vapor or gas to a liquid.

Conduction: Energy transferred directly from one object to another through contact.

Contract: To become drawn together or reduced; shrink.

Contrail: Water droplets or ice crystals condensing in the supersaturated air of the upper atmosphere as a result of an aircraft's passage.

Convection: The transfer of energy through circulation within a liquid or gas.

Counterclockwise: Opposite to the rotational direction of the hands of a clock.

Deflect: To turn aside from a course.

Density: The degree of compactness of a substance; mass/unit measure ~ weight.

Dew-point temperature: The temperature to which air must be cooled to reach the point of saturation of water vapor; the temperature at which dew forms.

El Niño: A band of warm ocean water temperatures that periodically develops off the western coast of South America. It can cause climatic changes such as floods and droughts in many regions of the world, particularly along the coasts of Chile, Peru, New Zealand, and Australia.

Ellipse: A closed, symmetric curve shaped like an oval.

Equilibrium: A state of balance between opposing forces or actions.

Equinox: The time when the sun crosses over the equator, making night and day approximately equal in length all over the earth.

Evaporate: To convert a liquid to a gas or vapor.

Evapotranspiration: The combination of water being evaporated and transpired.

Fluid: A substance, gas or liquid, that is capable of flowing.

Fossil fuel: Any naturally occurring carbon or hydrocarbon fuel, such as coal, petroleum, peat, and natural gas, formed by the decomposition of prehistoric organisms.

Front: A boundary at ground level between two different air masses as one air mass moves into the area of the other.

Grassroots: Used to describe organizations that begin and operate at the local or community level.

Greenhouse gases: Gases in the earth's atmosphere that absorb and emit radiation, or heat energy, in the thermal infrared range.

Hardiness: The ability of a plant to withstand cold temperatures.

Heat index: An addition to temperature data during hot and humid weather to indicate health risks more accurately than temperature alone.

Infrared radiation: A form of longwave radiation, or heat energy, given off by warm objects.

Isobar: A line on a weather map connecting points of equal barometric pressure.

Isoline: A line on a map or chart connecting points of equal value.

Isotherm: A line on a weather map connecting points of equal temperature.

Jet stream: Fast-flowing, narrow currents of air. The main jet streams are located near the tropopause.

La Niña: A cold phase in which the sea surface temperature across the equatorial eastern central Atlantic Ocean is lower than normal by 3–5°C. La Niña often follows El Niño.

Latent heat: Heat absorbed or radiated during a phase change.

Longwave radiation: Radiant energy emitted by lower energy objects, like a warm body giving off heat.

Mammatus: An ominous-looking cloud forming at the bottom of cumulonimbus clouds, with a sagging, pouch-like appearance.

Overwinter: To survive freezing conditions, often applied to perennial plants.

Precipitation: The amount of water that falls to earth as hail, drizzle, rain, sleet, or snow.

Radiant energy: Energy emitted in waves.

Relative humidity: The ratio, given as a percentage, of the amount of water vapor in the air compared to the amount of water vapor in saturated air.

Saturation: An atmospheric condition at which the moisture level is at 100% relative humidity.

Scale: A proportion that shows the relative size between objects but that is not the same size as the original objects.

Shortwave radiation: Radiant energy emitted by high-energy sources like the sun or a light bulb, which emit light energy.

Solar maximum: A time of great solar activity identified by the occurrence of many sunspots.

Solar minimum: A time of low solar activity identified by the occurrence of few sunspots.

Solstice: Either of two times a year when the sun is at its greatest distance north or south of the equator.

Sunspot: Dark, cool patches that appear on the surface of the sun.

Sunspot cycle: An approximately 11-year cycle of the occurrence of sunspots.

Thermal: A rising warm current of air caused by uneven heating of the surface.

Tropopause: The transition between the troposphere (where temperature decreases with altitude) and the stratosphere (where temperature increases with altitude).

Troposphere: The lowest portion of the earth's atmosphere.

Vortex: A whirling mass of water or air that forms a visible column or spiral.

Water vapor: Water in the form of an invisible, colorless gas.

Weather map: A daily report of observations and probable changes in the weather.

Weather station: One of 1,500 locations where the National Oceanic and Atmospheric Administration (NOAA) collects weather data.

Weather station model: A symbolic illustration showing weather conditions at a weather station recorded using abbreviations and symbols.

Windchill: Addition to temperature data during cold and windy weather to indicate health risks more accurately than temperature alone.

Zone: An area where the average annual minimum temperature is within a set temperature range.

PHOTO AND GRAPHIC CREDITS

WEATHER ACTIVITIES

Air Masses and Fronts worksheet • Page 125 • North American Air Masses • NOAA https;//www.noaa.gov/jetstream/synoptic/air-masses

Air Masses and Fronts worksheet • Page 126 • Cold front • Natalie Powell, Just Natalie Graphic Design, adapted from https://www.noaa.gov/jetstream/synoptic/air-masses

Air Masses and Fronts worksheet • Page 127 • Warm front • Natalie Powell, Just Natalie Graphic Design, adapted from http://www.noaa.gov/jetstream/synoptic/air-masses

Air Masses and Fronts worksheet • Page 128 • Cold front sketch • Adapted from https;//www.noaa.gov/jetstream/synoptic/air-masses

Pressure Systems worksheet • Page 137 • Air response to pressure • Ted Leuenberger

Pressure Systems worksheet • Page 138 • Air movement and Coriolis effect • Ted Leuenberger

Pressure Systems worksheet • Page 139 • High and low pressure systems and Typical US weather pattern • Courtesy NOAA, https://www.weather.gov/lmk/basic-fronts

Pressure Systems worksheet • Page 139 • Opt. • High and low pressure system • Courtesy NOAA https://www.noaa.gov/jetstream/synoptic/origin-of-wind#:~:text=So%20why%20does%20air%20 spiral,up%22%20at%20a%20given%20spot

Weather Station Models • Page 142 • Weather map • NWS https://www.weather.gov/forecastmaps

Weather Station Models • Page 143 • Weather station model • Purdue Extension Service

Weather Station Models worksheet • Page 144 • Common weather conditions • Purdue Extension Service

Weather Station Models worksheet • Page 145 • Wind symbols • Purdue Extension Service

Weather Station Models worksheet • Page 146 • Common weather symbols • Purdue Extension Service

Weather in the Troposphere worksheet • Page 152 • Wind patterns & Jet streams • Courtesy NOAA https://www.nesdis.noaa.gov/about/k-12-education/atmosphere/what-the-jet-stream and https://www.weather.gov/source/zhu/ZHU_Training_Page/winds/Wx_Terms/Flight_Environment.htm

Windchill and Heat Index worksheet • Page 159 • Windchill chart • Courtesy NOAA https://www.weather.gov/safety/cold-faqs

Windchill and Heat Index worksheet • Page 160 • Heat index chart • Courtesy NOAA https://data.usatoday.com/heat-index-forecast

Climate and Climographs worksheet • Pages 164, 168–169 • Climographs • Natalie Carroll

Drought Monitoring • Page 172 • Palmer Z Index conditions • Courtesy NOAA www.ncdc.noaa.gov/sotc/drought

Sun-Earth Relationship worksheet • Page 191 • Variations in temperature and CO_2 • Courtesy U.S. EPA https://archive.epa.gov/climatechange/kids/basics/past.html

Sun-Earth Relationship worksheet • Page 195 • Earth's orbit around the sun • Courtesy National Weather Service • https://www.weather.gov/cle/seasons

Sun-Earth Relationship worksheet • Page 197 • Deviation from annual mean temperature • Courtesy NASA http://data.giss.nasa.gov/gistemp/

Sunspot Cycle worksheet • Page 199 • Sunspot • Courtesy NASA https://www.weather.gov/fsd/sunspots

Sunspot Cycle worksheet • Page 200 • Deviation from annual mean temperature • Courtesy NASA http://data.giss.nasa.gov/gistemp/

Sunspot Cycle worksheet • Page 201 • Yearly averaged sunspot numbers • Courtesy NASA www.nasa.gov/mission_pages/sunearth/news/solarcycle-primer.html#.VAidE_ldWSo

Sunspot Cycle worksheet • Page 202 • Sunspot cycles • Courtesy National Weather Service https://www.weather.gov/fsd/sunspots

Sunspot Cycle worksheet • Page 203 • Mean annual temperature • Courtesy NASA www.earthobservatory.nasa.gov/IOTD/view.php?id=76975

NOTES

FACILITATOR'S GUIDE

Weather and Climate Science Education for Grades 3–12

NOTE

THANK YOU for taking the time to help a young person learn more about weather and climate science. Weather watching is useful on a daily basis and can even be a lifelong hobby. Learning about the earth's climate and climate change is important to understanding and responding to our changing weather.

Weather and Climate Science Education for Grades 3–12 is for youth who enjoy learning about science, especially weather and climate. Level 1 introduces basic terms and concepts for youth in grades 3–5. Activities focus on understanding the signs of weather. Level 2 activities introduce youth in grades 6–8 to more complex weather topics, understanding climate, and making and using weather instruments. Level 3 activities are divided into two major sections: weather and climate. Level 3 delves deeper into weather and climate science concepts to prepare youth to be well informed and to study these topics at a college or university. Youth are encouraged to supplement their learning by consulting knowledgeable people and current written materials in Level 3 with references from governmental and/or university sources (online extensions .gov and .edu).

This Facilitator's Guide answers questions in the youth manual and suggest ways to enhance the activities. It also offers additional information about working with youth.

NOTE TO THE PROJECT FACILITATOR

Thank you for helping students learn about weather and climate science as a project facilitator. This facilitator's guide answers the questions in the youth guides and suggests ways to facilitate the activities. It also gives additional information about working with youth, for those who wish to learn more.

You can help youth get the most out of this project by being enthusiastic and asking thoughtful questions. Team up with youth to help them select goals, identify resources, gain confidence, and evaluate their own progress. The activities are designed so youth have an opportunity to learn by doing with your help and guidance.

Your main role is to provide a safe, supportive environment for youth to practice important life skills as they explore the world of weather and climate science through many exciting activities.

RECOMMENDATIONS

- Gather materials. Youth should collect the materials listed in the tool kit before doing the activity. All the supplies used in these activities can be found around the home or purchased at minimal cost.
- Read and understand the activity. Each activity is self-contained.
- Encourage youth to try different activities and to enjoy learning.
- Show the relevance of each activity. Explore with youth the interconnectedness of each activity to the real world.
- Emphasize relying on others. Expand the learning activity and get additional supporting information from your Extension office. The Internet can provide a wealth of information to enrich learning. Take care to use sites that are supported by educational organizations (.edu), government agencies (.gov), professional societies, national organizations, and not-for-profit groups associated with the study of weather and climate.
- Keep safety in mind at all times. Some activities are done outdoors where wasps, bees, and other potentially harmful insects might be. Keep a watchful eye on the surroundings, and look out for weather changes.
- Visit Purdue University's Department of Earth, Atmospheric, and Planetary Sciences Teacher Resource website, http://www.eaps.purdue.edu/outreach/teacher_resources.html, for new and supporting educational materials.

LEARNING GOALS

Level 1

- Begin to learn about weather and climate science.
- Learn basic terms and concepts.

Level 2

- Understand drivers of weather systems and climate change.
- Develop further understanding of, and an appreciation for, weather and climate science.

Level 3

- Understand why learning about weather and climate science is important.
- Educate others about weather and climate science.
- Become a lifelong learner; continue to read and learn about weather and climate science.

KEY SCIENCE CONCEPTS

Weather is the state of the atmosphere at a particular place and time. An area's climate and many local factors—temperature, wind, cloudiness, precipitation, and relative humidity—influence the weather.

Climate is the usual, long-term weather conditions for an area, the average weather over time.

Climate system: How earth's water and gases flow or change state because of the sun's energy.

Climate change

"ThEarth's energy is balanced when the amount of energy coming in equals the amount going out. The temperatures we are used to will change when the energy input and outgo are not in balance. When the energy going out is greater than the incoming solar energy an ice age occurs. Similarly, when the incoming solar energy is greater than the energy leaving our atmosphere global warming occurs. Cloud cover is an important factor in this balance as clouds can block solar energy from coming in and block it from going out.

Climate change occurs when the earth's average surface temperature changes on a global scale. Although local average temperatures often fluctuate from year to year because of local weather systems, the global average usually stays about the same. But when the global average temperature begins to rise it is an indicator of climate change. The average temperature was 59°F in 2024, the warmest year on record.

Reference: Atmospheric Radiation Measurement Program (ARM), https://armgov.svcs.arm.gov/.

LIFE SKILLS

Programs help students develop knowledge and skills that help them become caring, competent adults. In "Targeting Life Skills in 4-H" (University of Florida IFAS Extension, 4-H FS 101.9) Norman and Jordan define life skills as competencies that assist people in functioning well in the environments in which they live. 4-H uses the Targeting Life Skills Model (Hendricks, 1998) to identify important assets that youth can learn through 4-H programming. The model uses the 4-H Pledge to categorize various life skills under four general competency areas: Head, Heart, Hands, and Health.

Head: knowledge, reasoning, and creative competencies

- Thinking—using one's mind to form ideas and make decisions; to imagine; to examine carefully in the mind; to consider
- Managing—using resources to accomplish a purpose

Heart: personal and social competencies

- Relating—establishing a mutual or reciprocal connection between two people that is wholesome and meaningful to both
- Caring—showing understanding, kindness, concern, and affection for others

Hands: vocational and citizenship competencies

- Giving—providing, supplying, or causing to happen; social responsibility
- Working—accomplishing something or earning pay to support oneself through physical or mental effort

Health: health and physical competencies

- Living—acting or behaving; the manner or style of daily life
- Being—living one's life; pursuing one's basic nature; involved in personal development

Targeted life skills are listed for each activity in this curriculum to remind youth that they are learning more than subject matter knowledge. Facilitators should review the life skills to understand the overarching goals of the activity with respect to youth development.

THE EXPERIENTIAL MODEL

The 4-H National Headquarters Fact Sheet explains, "Experiential learning takes place when a youth is involved in an activity, looks back at it critically, determines what was useful or important to remember, and uses this information to perform another activity. 4-H uses this hands-on learning approach to teach new topics and life skills." Experiential Learning Principles, www.4-h.org/resource-library/curriculum/development/.

The experiential learning model is a way of teaching to help youth make the most of any activity. It distinguishes 4-H activities from many other educational methods. Experiential learning is a process that allows youth to first learn by doing, before being told or shown how, and then to process the experience. Activities are designed so youth experience a learning activity; share what they did; process what they did through discussion, analysis, and reflection; generalize what they learned to test their comprehension and appreciation of the activity; and think about how they can apply what they learned to other situations.

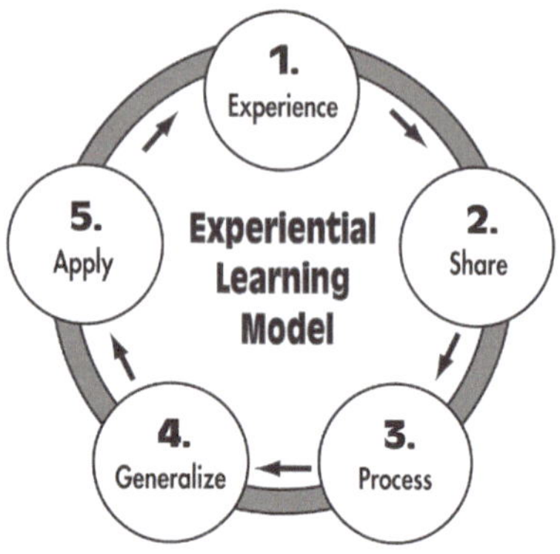

Using the experiential learning process in group settings offers these advantages:

- The adult can quickly assess the youth's knowledge of the subject.
- The youth builds on past experience or knowledge.
- The adult acts as coach rather than as teacher.
- The youth relate the experience to their own lives and experiences.
- Mentors may use a variety of methods to involve the youth in the experience.
- Youth with different learning styles can be successful.
- Discussions can move from the concrete to the abstract and analytical, which particularly benefits youth in middle school and high school.
- Youth are stimulated to learn through discovery and to draw meaning from the experience.
- Youth can work together, share information, provide explanations, and evaluate themselves and others.
- Youth take responsibility for their own learning.

Evaluating youth learning using a simple rubric such as the one shown can help 4-H volunteer leaders assess the effectiveness of their teaching methodology and youth interest. Evaluate each step of the experiential model by indicating what you think the 4-H members learned in a particular activity (your best guess). Work on improving any low scores.

Step	Excellent	Average	Minimal
Experience			
Share			
Process			
Generalize			
Apply			

YOUTH DEVELOPMENT STAGES

Understanding the physical, mental, social, and emotional development of youth will help you when working with the 4-H members in your club. No two youth develop at the same rate, and transitions are often gradual. Your teaching and involvement helps 4-H club members grow and mature, and makes 4-H a rewarding and fulfilling experience.

Activities at 4-H club meetings are not always as successful as you, the volunteer leader, hope and plan for. Sometimes youth talk among themselves rather than listening to you; no one comes to a scheduled field trip; or no one speaks up to answer your questions when you are trying to involve the youth in a discussion. If you are working with a broad age range, the activity might be too simple for the older youth and too difficult for the younger ones. This is challenging for the 4-H leader. Giving the older 4-H members leadership opportunities can be effective.

Youth of the same age can vary greatly in physical, mental, social, and emotional growth and interests. These differences are even more marked between age groups. Research provides some generalities that can help you understand how to plan activities for different age groups.

EARLY ELEMENTARY (MINI 4-H)

This is an active age, so keep these children busy. They are concrete thinkers and need to understand what you want them to do and how to do it. They are generally more interested in making something than in completing a project—process is more interesting than product. Youth in this age group tend to seek adult approval and depend on adults, although their peers' opinions are beginning to be important. They do best in small groups with set rules and rituals. Competition is inappropriate for this age group.

UPPER ELEMENTARY

This is also a physically active age, so hands-on activities work best. Youth in the upper elementary grades are still fairly concrete thinkers—things are black and white, right or wrong—but are beginning to think logically and symbolically. Because this age group has a strong need to feel accepted, an adult should evaluate each product, rather than encouraging competition among peers with only one winner. This age prefers to know how much they have improved against past efforts and how to improve in the future.

These youth are beginning to identify with peers but continue to value adult guidance. They are also beginning to discover the benefits of making other people happy, but more for the benefits to themselves rather than the benefit to others. They begin to take responsibility for their actions and to develop increased independence of thought, which might allow them to try new things. Letting this age group help in the decisions of the club helps them start learning about leadership.

MIDDLE SCHOOL

Middle school youth are beginning to move to more abstract thinking. Justice and equality are important to this age. Therefore, project judging may now be viewed in terms of what's fair as well as a reflection of self-worth. They prefer to find their own solutions, rather than adults giving them. Try to provide supervision without interference. Independent thinking and actions begins to emerge. Avoid comparing middle school youth with each other; compare performance with past accomplishments.

Junior volunteer organizations often are popular with teens toward the upper end of this age group, particularly with opportunities for developing leadership.

HIGH SCHOOL

Most high school-aged teens know their abilities, interests, and talents. They tend to be concerned with themselves and their peer group. While they can understand the feelings of others, they lean toward self-absorption, particularly in the earlier years of high school. Relationship skills are usually fairly well developed. Getting a driver's license increases both independence and dating, and acceptance by members of the opposite sex is important.

High school-aged youth begin to think about the future and make realistic plans. They enjoy career exploration and preparation. Their vocational goals influence the activities they select.

Projects requiring research and creativity give teens an opportunity to demonstrate how much they have learned and what they can accomplish. Teens set goals based on their personal needs and priorities and generally reject goals that others set for them.

As teens master abstract thinking, they might try new ideas in ways that confuse adults. Teens can generally initiate and complete tasks without supervision. A leader can help by arranging new experiences in areas of interest to teens but must be sure to allow them plenty of input. Assume the role of advisor/coach for independent workers, rather than teacher/lecturer. Club meetings, rituals, and uniforms do not generally appeal to this group. However, many teens enjoy looking back on their achievements in 4-H and appreciate special recognition for leadership activities. By the time they graduate from high school and begin college or a career, youth feel they have reached full maturity and expect to be treated as such.

POSITIVE IMPACT

These guidelines give only a brief overview of child and youth development as a resource to help you plan your activities as a volunteer leader. The publication Ages and Stages of Child and Youth Development* has more in-depth information and is available from your county Purdue Extension office.

You, as the club volunteer, are a valuable asset to your community and to your club's members. The guidelines for the stages of child and youth development, in combination with your special skills and interest in youth, will help you plan and carry out a successful 4-H program and make a positive impact on the lives of young people.

*Ages and Stages of Child and Youth Development, A Guide for 4-H Leaders, NCR-292, available at Purdue Extension's Education Store, www.edustore.purdue.edu; search NCR-292.

ACTIVITY FACILITATION AND ANSWERS: LEVEL 1

COMPARING CLIMATES

Big Picture: People often think of the different seasons as being different climates. An area's climate includes the seasons there. Different seasons are therefore part of our climate. Climates are classified in different ways. This activity highlights the temperate, tropical, desert, and tundra climates because these can be illustrated using regions in the United States. In this scheme, rainfall and temperature are the major components of climate. Other systems involve biomes or geologic features. Indiana is in the temperate climate. The word "temperate" means moderate, and the youth will discover that Indiana does not have the extreme differences in rainfall or temperature of other climates. Indiana's climate, even though it has great variation of weather, is moderate when considering extremes in temperature and rainfall.

FACILITATING THE ACTIVITY

Youth might be able to complete this activity without adult assistance. You should check that they are interpreting the data correctly.

Climate Answers:

Question: Look up the word "temperate" in the dictionary and write the definition.

ANSWER:
The definition probably includes the word "moderate" or something about not having extremes.

Question: The questions you answered from the data were extremes (highest and lowest of the averages). How many times did the temperate climate have extreme data in a category?

ANSWER:
Only 1

SHARE AND APPLY QUESTIONS

ANSWER:
Interpretive—no right or wrong answers

The highest average temperature?	Tropical	What is it?	75.9°F
The lowest average temperature	Tundra	What is it?	26.9°F
The highest July temperature?	Desert	What is it?	93.5°F
The lowest July temperature?	Tundra	What is it?	26.9°F
The highest January temperature?	Tropical	What is it?	67.2°F
The lowest January temperature?	Tundra	What is it?	-10.1°F
The highest average precipitation?	Tropical	What is it?	55.9 in.
The lowest average precipitation?	Desert	What is it?	7.7 in.
The highest July precipitation?	Tropical	What is it?	5.7 in.
The lowest July precipitation?	Desert	What is it?	0.8 in.
The highest January precipitation?	Temperate	What is it?	2.3 in.
The lowest January precipitation?	Tundra	What is it?	0.5 in.

GENERALIZE TO YOUR LIFE

Question: Why doesn't Indiana have many extremes in temperature and rainfall?

ANSWER:
Indiana has a temperate climate in which rainfall and temperatures are moderate.

Question: Temperature and rainfall are the major factors in a climate. What might some other differences be?

ANSWER:
Ideas could include vegetation, humidity, elevation, and others.

CONNECTIONS

There are many climates around the world. Encourage youth to investigate them and compare them to our climate.

There are different terminology systems. Some are based on temperature and precipitation, some on biomes (flora and fauna), and some on geologic components.

FLY HIGHER

Calculate the difference in average temperature and rainfall between January and July for each climate. What climate has the greatest extremes for temperature and rainfall?

The tropical climate has the smallest temperature difference, and the desert climate has the smallest precipitation difference (and the least precipitation overall).

LIFE SKILLS

- Weather Skills: Understanding that climates are different in other areas, and that Indiana has a temperate climate that does not have extremes in rainfall or temperature.
- Indiana Science Standards: 2.2.4
- NGSS: ESS2.D, 3ESS2-1, 3-ESS2-2
- Success Indicator: Youth understand that Indiana is in a temperate climate that includes changing seasons with moderate temperature and rainfall.

SOURCES

- Climate: www.climate-zone.com/climate/united-states
- Temperate image: www.nhptv.org/natureworks/nwep8c.htm
- Desert image: www.city-data.com/picfilesv/picv27841.php
- Tropical image: www.nps.gov/ever/parknews/upload/greategret.jpg
- Tundra image: www.animalcorner.co.uk/biomes/tundra.html

Climate/State	Indicator	January	July	Difference
Temperate/Central Indiana	Average Temperature	25.5°F	75.4°F	49.9°F
	Average Precipitation	2.3"	4.5 in.	2.2 in.
Desert/Central Arizona	Average Temperature	53.6°F	93.5°F	39.9°F
	Average Precipitation	0.7 in.	0.8 in.	0.1 in.
Tropical/South Florida	Average Temperature	67.2°F	82.6°F	15.4°F
	Average Precipitation	2.0 in.	5.7 in.	3.7 in.
Tundra/Central Alaska	Average Temperature	-10.1°F	62.5°F	72.6°F
	Average Precipitation	0.5 in.	1.9 in.	1.4 in.

Note that the tundra climate has the largest average temperature difference, and the tropical climate has the largest average precipitation difference.

COUNTRY OF COLORS

Big Picture: Weather is important to gardeners and farmers. The climate we live in is one of the factors that determine what can be grown and when it should be planted. Hardiness zones are based on the average minimum temperatures for each zone. Lower numbers indicate the most northerly area where plants will survive the winter, and higher numbers are the most southern area where they will thrive. The Hardiness Zone Map should be used as a guideline only, because local climates can vary and microclimates occur in some areas. Factors such as rainfall, wind, humidity, soil type, altitude, sunlight, and cultural practices also affect plant survival.

ANSWERS:
The Indiana Hardiness Zones map shows the three hardiness zones for the state. Youth do not need to be precise when drawing the boundaries; rather, they should understand where the general zones are and see that they are not based on latitude alone.

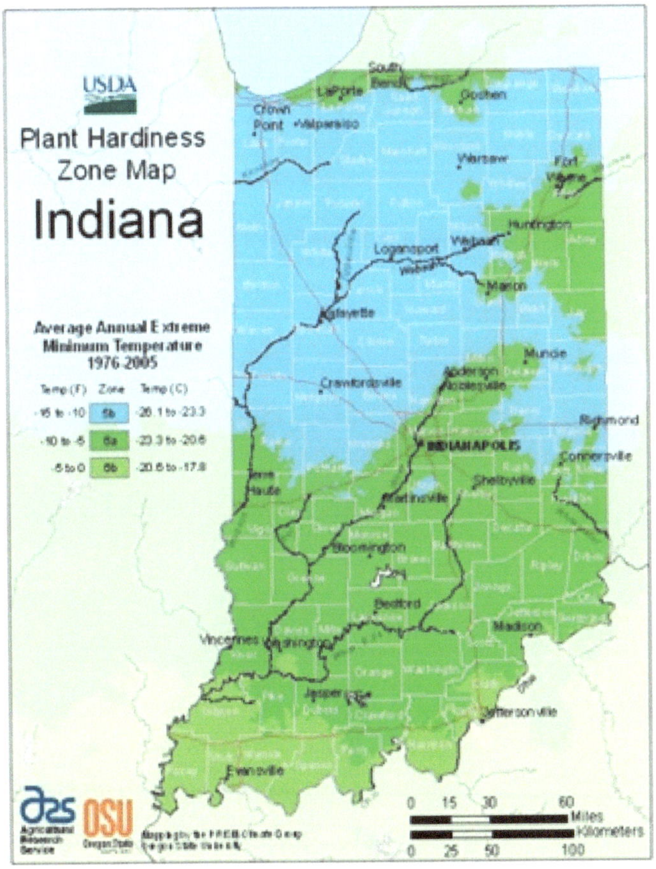

FACILITATING THE ACTIVITY

Youth should be able to complete this activity on their own. When reviewing the experiential learning questions (Chat), encourage youth to think about how the values differ for different parts of the state, and how they might differ for other parts of the U.S. Most youth enjoy using the interactive USDA Hardiness Zone Map (see "Fly Higher").

Chat SHARE WHAT HAPPENED

Question: How many zones did you find in Indiana?

ANSWER:
Indiana has two hardiness zones: Zone 5 (-20 to -10°F) and Zone 6 (-10 to 0°F)

APPLY

Question: What is the average annual extreme minimum temperature range where you live?

ANSWER:
Northern Indiana: -20 to -10°F; Southern Indiana: -10 to -0°F

Question: What do you think would happen if you planted a hardy perennial flower, a plant that lives more than one year, in Hardiness Zone 9, which includes the warmest parts of the United States?

ANSWER:
The flower would probably not survive the Indiana winter so would need to be planted each spring (annual).

GENERALIZE TO YOUR LIFE

Question: How do the hardiness zones affect what farmers grow?

ANSWER:
Like gardeners, farmers must plant cold-sensitive crops when the temperatures are warm enough to support them. Corn and soybeans are commonly grown in Indiana.

FLY HIGHER

Question: Where are the warmest zones in the U.S.? What is their average minimum temperature?

ANSWER:
The warmest zones are in the most southern reaches of the U.S. (Florida, Texas) and along the Pacific coast (California, Arizona). The average minimum temperature is 30–40°F /40--0°F.

Question: Where are the coldest zones in the U.S.? What is their average minimum temperature?

ANSWER:
The coldest zones are along the U.S.-Canadian border. Maine, Vermont, New Hampshire, New York, Michigan, Wisconsin, Minnesota, North Dakota, Montana, and Idaho all have areas classified as Zone 3, as do Wyoming and Colorado.

LIFE SKILLS

- Weather Skills: Learning how gardeners and farmers use temperature averages.
- Indiana Science Standards: 2.2.5
- NGSS: ESS2.D
- Success Indicator: Youth can describe the effect of last frost dates on gardeners.

SOURCES

USDA Plant Hardiness Zone Map, http://planthardiness.ars.usda.gov/PHZMWeb/

DEFINING WEATHER WORDS

Big Picture: This activity explores weather words that youth already know and helps them learn definitions of words they don't know. As they continue to study the 4-H Weather and Climate Science manuals, their vocabulary will increase, and they will be able to read or talk to others about the weather and gain greater understanding.

ANSWER:
The four H's in "4-H" stand for Head, Heart, Hands, and Health.

FACILITATING THE ACTIVITY

Encourage youth to create their own definitions. While some answers might be closer to the dictionary definitions than others, the important thing is for youth to think about each term and how they would describe them. Youth in grades 3–5 have little experience writing definitions, and it can be difficult for anyone to define terms that we use daily. This activity is designed to make the student think about how they might describe weather words.

Scrambled Letters	Weather Word	Dictionary Definition
ahrewet	weather	The state of the atmosphere with respect to temperature, wind, cloudiness, moisture, pressure, etc.
rai	air	A mixture of odorless, tasteless gases such as nitrogen and oxygen that surround the earth and form its atmosphere
tarew	water	A transparent, odorless, tasteless liquid, a compound of hydrogen and oxygen (H_2O)
athe	heat	A condition of being hot or warming
nsu	sun	Star around which the earth and other planets revolve and from which they receive light and heat
herta	earth	The third planet from the sun, the planet we live on
nwid	wind	Air in natural motion, moving horizontally at any velocity along the earth's surface
nria	rain	Water that falls to the earth in drops with a diameter of more than 0.5 mm
lhia	hail	Showery precipitation in the form of irregular pellets or balls of ice more than 5mm in diameter falling from a cumulonimbus cloud
scoldu	clouds	A visible collection of particles of liquid water or ice suspended in the air
ateuprmeet	temperature	Degree of hotness or coldness measured on a definite scale
etamilc	climate	The average condition of the weather at a place, over a period of years, as exhibited by temperature, wind velocity, and precipitation
miudtihy	humidity	The amount of water vapor present in the air
nswo	snow	Precipitation that falls in the form of ice crystals

SHARE AND GENERALIZE QUESTIONS

ANSWER:
Interpretive—no right or wrong answers

APPLY

Question: How does knowing weather words help you talk about the weather?

ANSWER:
Understanding the terminology and basic concepts in any discipline, hobby, or other interest is important so youth can understand written and oral communications and accurately communicate their thoughts and ideas.

CONNECTIONS

We use a variety of terms and definitions in all walks of life. Many of these terms might not be familiar to someone outside of a specific discipline. Youth might enjoy talking about the terms and definitions they use in 4-H or a favorite sport. Learning the jargon (technical terminology) of a specific discipline facilitates discussion among group members and is one of the first steps for youth to start learning about weather and climate. If youth continue to study weather topics, their weather vocabulary will increase.

LIFE SKILLS

- Weather Skills: Learning weather terminology.
- Indiana Science Standards: 2.2.1
- NGSS: ESS2.D
- Success Indicator: Youth can describe different weather terms.

EARTH'S SURFACES

Big Picture: Sunlight has very little warming effect on the air. Sunlight travels through the air. We can see for miles through the air because it does not absorb light energy. The sun warms the earth, and the earth warms the air. Certain surfaces, like snow and ice, reflect most of the sunlight back into space. Dark surfaces like sand, rocks, and soil absorb sunlight efficiently, so the air above these surfaces warms more quickly.

Water is the exception. Bodies of open water absorb sunlight better than most of the surfaces on earth. This absorption occurs into the depths of the water, so water has less effect on the air above it. Open water absorbs sunlight well but releases heat energy slowly. Manmade surfaces also affect how air is warmed. Large urban areas become "heat islands" and can have definite effects on the weather. Areas like plowed fields can also change the dynamics of heating.

In this activity youth discover that darker surfaces heat up faster than lighter surfaces. They make the connection that different surfaces on earth heat the air above them at different rates, causing air movement.

FACILITATING THE ACTIVITY

This activity should require little assistance. Finding the right thermometer is important. Most standard outdoor thermometers have a protected bulb. This activity works best with a thermometer with an exposed bulb, as shown in the example. These thermometers are inexpensive and can be found in science catalogs online. Your school's science teacher probably has many of these and would be happy to loan one out.

Question: Does the type of light bulb (incandescent, fluorescent) matter?

ANSWER:
Technically not. A flashlight will work. This type of activity can be tricky. As long as the thermometer bulb is in contact with the paper, any light source will work fine. The light energy is being tested. Incandescent lights emit a lot of infrared (heat) energy, which increases the odds of good results but is kind of like cheating. A thermometer that creates an air space between the paper and the bulb changes the heating dynamics from simple conduction between the paper and bulb to a complicated energy transfer system that involves heating the air around the bulb, which uses

convection to move heat energy away and pull cooler air under the paper.

SHARED WHAT HAPPENED

Question: Which color did you think would heat the most?

ANSWER:
Interpretive—no right or wrong answer

Question: Which color heated the most?

ANSWER:
Black should heat up the most.

Question: What surfaces on earth would be like each of the colors used in your experiment?

ANSWER:
White is like snow and ice; black is similar to pavement; brown represents tilled or bare ground; green is similar to grass; and blue is like oceans, lakes, and streams. (However, blue does not correctly reflect water's ability to absorb energy. And water is not blue. Although water is actually the most effective absorber, the heating is not at the surface; it continues into the depths, leaving the surface cool.)

Question: Which earth surfaces would heat the air above them the most? The least?

ANSWER:
Most: deserts, roads, or any darker surface. Least: snow and ice.

APPLY

Question: Why does the sun's uneven heating of ocean and land surfaces cause breezes?

ANSWER:
A sandy beach on a hot, sunny day is hot to the touch, yet the water nearby feels cool and refreshing. These differences cause the air above the beach to become warmer and rise.

Question: Why is the air over the land rising?

ANSWER:
The air over the land has become warmer. Because it is now lighter than the denser air that surrounds it, it rises.

Question: What is causing the sea breeze?

ANSWER:
The cooler air over the water is pulled in to replace the rising warm air.

Question: How does the surface affect breezes?

ANSWER:
Uneven heating of the surfaces gives the air above them different temperatures, causing air movement.

GENERALIZE TO YOUR LIFE

Question: How might the different surfaces affect the wind in each place?

ANSWER:
Some locations have surfaces that heat unevenly, causing air movement.

Large parking lots and built environments with no trees are warmer than forested areas.

Question: How might humans change the earth's surfaces and affect the local weather?

ANSWER:
Answers should include positive and negative examples: building roads, parking lots and buildings; destroying forests or restoring forests. Encourage positive suggestions like planting trees and creating parks and nature preserves.

CONNECTIONS

Spending time at the beach is fun. One reason people love the beach is because they can enjoy the sun and still stay cool, as the natural heating of the surfaces provides a constant cool breeze from the ocean. Even though the ocean absorbs heat energy from the sun, that energy is distributed in the depths of the water, and the surface remains cooler. Land absorbs the sun's energy at the surface, but little of the heat energy moves through the sand or soil. Youth can dig a few inches into beach sand to find cooler temperatures. Deep soil maintains a more constant temperature. This is the basis for geothermal heating and cooling systems.

A striking example of differences in surfaces can be seen in Australia, where a 2,000-mile fence was built to keep rabbits out of agricultural crops. This created a large area of farmland next to thick native vegetation. The result was an atmospheric boundary with clouds forming on one side and not the other.

LIFE SKILLS

- Weather Skills: Understanding that sunlight warms surfaces in different ways, which causes differences in the air above these surfaces.
- Indiana Science Standards: 2.2.2, 2.2.5, 4.2.6
- NGSS: ESS2.D
- Success Indicators: Youth record different temperatures and understand that darker surfaces heat the air above them more on a sunny day than lighter surfaces do; and that these differences can influence the weather.

H_2O

Big Picture: Youth should understand that water vapor condenses out of air when the air is cooled. More water vapor can exist in warmer air than in cooler air. Cold air cannot hold water vapor very well. For air to cool in the atmosphere, it must rise. When air rises, the air pressure decreases, which causes cooling. Using a spray can causes it to cool as the pressure is released. If you let air out of a tire with your finger, your finger gets cold as the pressure is released. The air in an existing cloud is rising; otherwise, the cloud would not exist. With the right conditions, a cloud begins to produce precipitation, and the cycle begins again.

On a sunny day clouds form cotton ball shapes in the sky. The sun heating the ground and making the air rise causes these. When the sun sets in the evening, they disappear because the rising of air stops.

FACILITATING THE ACTIVITY

Youth might be able to complete this activity without adult assistance. It is difficult to understand that the water forming on the jar comes out of the air. Many youth conclude that the water came from the ice or from inside the jar. This phenomenon is called sweating. "The glass is sweating," implies that water is going through the glass. Youth might need to be directed to the idea that the water existed as water vapor in the air before condensing on the jar.

SHARE WHAT HAPPENED PART 1

Question: Which jar had water condensing on the outside?

ANSWER:
Water condensed on the outside of the jar with ice in it.

Question: Do you think that condensed water came from water vapor in the air?

ANSWER:
Yes, the water condensed from water vapor in the air.

Question: Why did water condense only on the cold jar?

ANSWER:
The air must be cooled to get water vapor to condense.

SHARE WHAT HAPPENED PART 2

Question: What did you observe at each step on the inside of the jar?

ANSWER:
After shaking the water in the jar, the glass sides were clear. After pouring ice water over the jar, the inside of the glass fogged up.

Question: What happens to the temperature of the air when water condenses?

ANSWER:
Whenever the air was cooled, water vapor condensed.

APPLY

Question: Study the picture of the water cycle. How does what you learned about precipitation relate to the water cycle?

ANSWER:
Water evaporates into water vapor at the surface and condenses into clouds up in the air. Rain from the clouds brings the water back to the surface.

GENERALIZE TO YOUR LIFE

Question: When does dew form? When does frost form?

ANSWER:
Dew forms in the morning. Frost forms in the evening.

Question: Would you ever expect dew or frost to form when things are warming up?

ANSWER:
No, the air must cool for water to condense.

Question: Have you ever watched clouds on a sunny day? What do they look like?

ANSWER:
A variety of answers may be noted: marshmallows, cotton balls, puffy, etc.

CONNECTIONS

Many things in our lives depend on evaporation and condensation. Frost-free freezers work by driving cold, dry air through the freezer to evaporate frost. Air conditioners produce a lot of water because cooling the air causes water vapor to condense. Dehumidifiers condense water vapor out of the air to make it more comfortable in the summer, and humidifiers evaporate water into the air to make it more comfortable in the winter.

LIFE SKILLS

- Weather Skills: Understanding how temperature and cloud formation are related and contribute to the water cycle.
- Indiana Science Standards: 2.1.1
- NGSS: ESS2.D
- Success Indicator: Youth understand how evaporation and condensation are a crucial part of the weather cycle.

INVISIBLE AIR

Big Picture: Air is an essential part of our lives and all around us, even if we are not aware of its presence. Air is often taken for granted, with little thought given to how and why it moves. In this activity youth observe the expansion and contraction of air as well as the buoyant effects of these changes. Changes in air temperatures create wind patterns and movement of air all over the earth.

Youth learn that air takes up space by seeing the air taking up the space in a glass and pushing the water level up in the bowl. Ice cools the air in the testing jar, causing the air to contract. The contraction pulls air from outside the jar into the straw, making bubbles in the water. As the air is cooled, molecules lose energy and move closer, causing the air to contract. Placing hands on the jar warms the air. As the air warms, the molecules gain energy and move farther apart, causing the air to expand. The expanding air forces the water up the straw.

In the balloon activity, youth see that as air contracts in the balloon, it becomes heavier than the other balloon. The mass of the balloon does not change; the air molecules only contract. The balloon is getting denser as the molecules lose energy and move closer together. This causes the balloon to become heavier (denser) than an equal volume of the warmer air around it, which causes that air to sink. The difference between weight and density is a complex idea not commonly understood by youth at this age. Youth should start to observe moving air (wind) as a result of the warming and cooling of air and the changes in the amount of space air takes up as temperature changes.

FACILITATING THE ACTIVITY

Making the jar apparatus—drilling the metal lid and using the glue gun — requires adult supervision. For increased safety, put a block of wood under the lid when drilling. The size of jar does not matter as long as it is glass and the lid and straw seal tightly. Coloring the water makes it easier to see it in the straw, but do not make the color too dark; one drop is usually more than enough unless you use yellow.

PART 3 QUESTIONS

Question: Which is heavier (denser), warm or cold air?

ANSWER:
Cold air

Question: If the air in a balloon is cooled, will the balloon's weight (density) change?

ANSWER:
Yes, the air becomes heavier (denser).

SHARE WHAT HAPPENED PART 1

Question: What happened to the water in the bowl when you inserted the upside-down glass?

ANSWER:
Pushing the glass into the bowl causes the water level in the bowl to increase.

Question: What kept the water from entering the glass?

ANSWER:
The air is taking up that space.

SHARE WHAT HAPPENED PART 2

Question: What did you observe when you cooled the air in the jar with the straw?

ANSWER:
Air is being pulled through the straw into the jar.

Question: What did you observe when you warmed the air in the jar with the straw?

ANSWER:
The expanding air in the jar pushes the water up the straw.

SHARE WHAT HAPPENED PART 3

Question: Did the balloons on the yardstick stay balanced when you cooled one of the balloons?

ANSWER:
No, the cooled balloon sank.

APPLY

Question: Why does cooled air sink?

ANSWER:
Cooled air sinks because it contracts and becomes heavier (denser) than the air around it.

Question: Why does warmed air rise?

ANSWER:
Warmed air rises because it expands and becomes lighter (less dense) than the air around it.

GENERALIZE TO YOUR LIFE

Question: Is there a breeze outside today? How might changes in the air temperature have caused that breeze?

ANSWER:
As air warms, it rises and causes movement of the air around it (a breeze).

Question: Have you ever seen a hot air balloon? How do you think a hot air balloon works?

ANSWER:
The air in the balloon expands as it is heated. The hot air becomes lighter (less dense) than the air around it, causing the balloon to lift.

LIFE SKILLS

- Weather Skills: Understand the importance of air and its relationship to weather, especially wind.
- Indiana Science Standards: 2.2.2

- NGSS: PS1.A: Gases are made of particles too small to see and that are moving freely around in space.
- Success Indicators: Youth understand that air takes up space. It contracts when cooled and expands when heated, which greatly affects wind patterns and, in turn, weather.

CONNECTIONS

The uneven heating of the earth's surface, and the resulting changes in air density, is the driving force of all wind and weather events. These dynamics also result in heat energy moving from warm areas to cool areas.

READING ABOUT WILD WEATHER

Big Picture: Reading about weather events is a good way to get involved with weather in a more personal way. Each of these books relates real stories about people who deal with weather events and the scientists who study these events. There are four books to choose from, on tornadoes, lightning, floods, and blizzards.

FACILITATING THE ACTIVITY

There are several possible sources for these books. Schools might have them as part of their leveled reading library; they are part of Scholastic's Hello Reader! Science series; and they can be found online at reasonable prices. Each book is similarly formatted with five chapters in about 46 pages. Encourage youth to share personal experiences as well as reported events.

SHARE WHAT HAPPENED

Question: What was the most interesting thing you learned in each of the five chapters?

ANSWER:
Interpretive—no right or wrong answers

Question: Each Wild Weather book introduces a scientist who studies a weather event. Who was the scientist you read about, and what kind of work does he or she do?

ANSWERS:
Lightning: Maribeth Stolzenburg launches weather balloons into thunderstorms.

Floods: Kate White studies ice jams that might cause flooding.

Tornadoes: Tim Marshall is a tornado chaser.

Blizzards: John Cortinas studies past storms to help predict new storms.

APPLY AND GENERALIZE QUESTIONS

ANSWER:
Interpretive – no right or wrong answers

LIFE SKILLS

- Weather Skills: Learning that anecdotal experiences can provide insight into current ideas and research on severe weather events.
- Indiana Standards: 2.2.4
- NGSS: ESS2.D
- Success Indicator: Youth become familiar with different types of severe weather events and the preparations for dealing with them.

'TIS THE SEASON

Big Picture: Seasonal changes are the environment's visible reaction to temperature and precipitation. Youth should understand that while weather varies from day to day and month to month, the changes are generally predictable. Seasons are weather patterns that describe weather over a period of time (months). In this activity youth examine something that affects their daily life but is part of a weather pattern larger than the day's weather.

FACILITATING THE ACTIVITY

Encourage youth to be creative in thinking of something that represents each season for them.

(Note: there are no right or wrong answers and no good, better, best response.) Youth have varying degrees of artistic skill as well as understanding about the separate seasons. It is important to encourage them to use their own strengths, not to compare themselves to others.

If youth are having trouble thinking about how to show the different seasons, you might ask them how trees look in the different seasons; how they dress; or what outside sports and activities they like to do in January, April, July, and October.

SHARE AND APPLY QUESTIONS

ANSWER:
Interpretiveno right or wrong answers

GENERALIZE TO YOUR LIFE

Question: How does the season affect what you wear?

ANSWER:
Youth most often mention temperature and precipitation. They might also think about windchill (winter), humidity (summer), and cloud cover when deciding what to wear. For example, they might like to wear sunglasses when it is sunny outside.

Question: How does the season affect what you do outside?

ANSWER:
Youth may discuss sports (baseball, soccer, swimming, tennis, golf, skiing, skating, etc.); outdoor 4-H activities (riding, walking show animals, collecting leaves or fossils, photography, etc.); or whether they ride a bike or in a convertible with the top down.

Question: How does the season affect plants?

ANSWER:
Youth may discuss watering the garden, trees that lose their leaves, when they must mow the grass, or other plant growth affected by the seasons.

CONNECTIONS

Seasons dictate what we do daily and how we make plans for the future. Youth may not think about how certain activities are planned around seasons: sports, school, gardens, even the winter and summer Olympics! Youth might prefer to show the effects of seasonal weather on a single object, like a tree, or show a holiday that occurs during each season.

LIFE SKILLS

- Weather Skills: Identifying different types of weather that correspond with each season.
- Indiana Science Standards: 2.2.4
- NGSS: ESS2.D: 3-ESS2-1:
- Success Indicators: Youth can identify differences between seasons and how these differences change personal behaviors.

WATCHING THE WIND

Big Picture: The Beaufort Wind Scale has long been used as a reliable way to estimate wind speed. Wind speeds on this scale are organized into categories based on general strength.

FACILITATING THE ACTIVITY

Ensure that each youth has a reliable local source for local wind speed.

SHARE WHAT HAPPENED

Question: Was it to difficult use the Beaufort Wind Scale?

ANSWER:
Interpretive—no right or wrong answer

Question: How did your readings compare to the actual wind speeds reported from your weather source?

ANSWER:
Results will vary depending on equipment, accuracy, and local weather.

Question: Do you think the Beaufort Wind Scale is useful?

ANSWER:
Interpretive – no right or wrong answer

APPLY

Question: What are some other objects affected by the wind that could help you estimate wind speed?

ANSWER:
Observations could include flowers, bushes, dust from a road, flags, awnings, clothes on a clothesline, or other items that can be blown about by the wind.

GENERALIZE TO YOUR LIFE

Question: When would it be useful for you to estimate wind speed? Are there particular places or situations where this knowledge might come in handy?

ANSWER:
Estimating wind speed is useful when an accurate instrument or media source is not available. It also encourages youth to observe and interpret weather conditions.

LIFE SKILLS

- Weather Skills: Estimating wind speed using the Beaufort Wind Scale.
- Indiana Standards: 2.2.1, 2.2.2, 2.2.3
- NGSS: ESS2.D, 3-ESS2-1
- Success Indicator: Youth observe and compare estimated wind speeds and reported wind speeds.

SUPPLEMENTAL INFORMATION AND RESOURCES

This activity was adapted from http://www.spc.noaa.gov/faq/tornado/beaufort.html.

WEATHER AFFECTS PLANS

Big Picture: Keeping weather data is a lifelong hobby for some people, particularly gardeners and farmers, because their crops depend on weather conditions. Youth keep temperature data from professional sources and consider how the temperature affects their outside activities.

FACILITATING THE ACTIVITY

Youth might need help obtaining the daily high and low temperatures. Radio stations often give the expected values, as does the Weather Channel. If youth have Internet access, they can enter their city and state or zip code at the NOAA (National Oceanic and Atmospheric Administration) website, http://forecast.weather.gov/, and get temperature data. Encourage youth who enjoy keeping records to make their own charts. You can help youth calculate the difference in daily high and low temperatures and discuss when these values tend to be larger (spring and fall), or smaller (summer and winter), and how this affects us.

SHARE WHAT HAPPENED

ANSWER:
Interpretive—no right or wrong answers

APPLY

Question: Did the weather affect your outdoor activities?

ANSWER:
Interpretive—no right or wrong answer

Question: When is the weather most likely to affect your outdoor activities?

ANSWER:
Winter weather and rain generally affect outdoor activities the most.

Question: When has the weather affected family plans?

ANSWER:
Interpretive—no right or wrong answers

GENERALIZE TO YOUR LIFE

Question: What affect does weather have on:

ANSWER:
Vacation plans—Some people like to vacation where it is warm and sunny, so they vacation in the summer or in warm places during the winter. Some people like to snow ski, snowboard, or ice skate, so they vacation where there is ice and snow.

Building a house—Average temperatures affect the presence and/or size of heating and air-conditioning systems. Humid areas may use dehumidifiers; dry areas may need humidifiers. Homes in areas of the United States with cold winter temperatures need more insulation and windows that resist heat loss. People may choose to install fewer windows in cold areas to reduce heat loss.

Walking a trail or hiking—Expected temperatures and precipitation determine the clothing that youth wear or take on a walk or hike.

FLY HIGHER

Youth can collect weather data from local media outlets or on the NOAA website, www.noaa.gov, by entering their zip code (or city and state) in the search box on the left side of the homepage. They need help to study the graphs found as follows:

- Click on the graph under Hourly Weather Graph on the lower right of the forecast page.
- Study the graphs:
- Heat Index/Dew point/Temperature
- Wind/Gusts
- Relative Humidity/Precipitation Potential/Sky cover
- Thunder
- Rain (Note: The bar shows the likelihood of thunder; totals are given for various time spans across the bottom of the graph.)

LIFE SKILLS

- Weather Skills: Collecting data.
- Indiana Standards: 2.2.3, 2.2.4
- NGSS: ESS2.D, 3-ESS2-1
- Success Indicators: Youth understand how weather affects what we wear and what we do.

WEATHER ALERTS

Big Picture: Only the National Weather Service makes official weather announcements. These alerts are sent to local radio and television stations. You can hear the alerts earlier by having an NOAA weather radio in the home. Warnings are the most important alerts because they are issued when weather conditions are severe and imminent. Watches are issued when severe weather conditions are possible but not yet a reality. Advisories are for less severe types of weather.

FACILITATING THE ACTIVITY

Youth might need help confirming definitions for the different types of warnings. The following NOAA website has the specific definitions: www.erh.noaa.gov/lwx/Defined/

SHARE WHAT HAPPENED

Question: What are the three NWS alerts?

ANSWER:
Warnings, watches, and advisories

q	u	i	t	j	h	u	h	f	r	l	p
a	f	u	h	k	g	y	i	e	t	o	z
z	o	i	u	b	l	i	z	z	a	r	d
w	g	k	n	l	f	c	o	v	g	t	x
s	n	l	d	p	d	e	p	w	b	i	c
x	b	o	e	o	s	t	a	c	i	k	v
c	g	p	r	i	a	r	z	d	n	n	b
f	l	a	s	h	f	l	o	o	d	m	d
d	t	j	t	u	q	w	x	e	h	j	n
e	r	v	o	t	o	r	n	a	d	o	m
r	f	w	r	y	s	e	s	w	y	u	l
f	v	e	m	z	w	i	n	t	e	r	k

Question: Which alert means that you need to take immediate action?

ANSWER:
Warning

APPLY

Question: Have you recently heard or seen a warning, watch, or advisory? If so, what weather event was it for?

ANSWER:
Interpretive—no right or wrong answer

Question: Have you ever experienced a warning when you took action? If so, what action did you take?

ANSWER:
Interpretive—no right or wrong answer

GENERALIZE TO YOUR LIFE

Question: Does your family have a plan or preparations made for storm warnings? If so, what are the plans or preparations?

ANSWER:
Interpretive—no right or wrong answer

Question: Why might a family have a NOAA (National Oceanic and Atmospheric Administration) Public Alert Weather Radio?

Weather Alerts Data Sheet	Weather
Winds of 58 mph or higher with hail 1 inch in diameter or more and dangerous lightning.	Thunderstorm
Winds of 40 mph or higher, or wind gusts of 58 mph or higher, for one hour or more.	Wind
Imminent danger of rapid flooding in low-lying areas (two words).	Flash Flood
Blowing snow reducing visibility to a quarter-mile or less for three hours or longer, and sustained winds of 35 mph or greater or frequent gusts to 35 mph or greater.	Blizzard
A tornado is imminent.	Tornado
1/4 inch or more of ice accumulation.	Ice
The heat index value is expected to reach or exceed 110 degrees within the next 12–24 hours.	Heat
Reduced visibility to a quarter-mile or less over a large area for an extended period of time (two or more hours).	Fog
A significant combination of hazardous winter weather is occurring or imminent.	Winter

ANSWER:
These radios allow you to quickly learn about dangerous weather conditions. They can be set to broadcast weather alerts when they occur for any U.S. area. They are available with a battery and hand-crank backup.

FLY HIGHER

Create a plan for your family so that you are prepared for a tornado warning and and/or a winter storm warning.

Youth can get help for creating a plan by finding sources on the Internet or calling their county's emergency management office.

LIFE SKILLS

- Weather Skills: Recognizing the difference between warnings, watches, and advisories.
- Indiana standards: 2.2.6
- NGSS: ESS2.D
- Success Indicator: Youth understand how to react to the different weather alerts issued by the NWS.

SUPPLEMENTAL INFORMATION AND RESOURCES

Definitions for watches, warnings, and advisories: www.weather.gov/lwx/warningsdefined.

WEATHER OR CLIMATE?

Big Picture: The difference between weather and climate can be confusing. Climate includes all the different types of weather that happen year round and is based on the last 30 years of weather records. Weather is simply what happens at a specific time in the atmosphere. Therefore the temperature today is the weather; the average temperature for today is the climate.

Weather can vary widely in Indiana because of our changing seasons, but the seasons are part of our climate. Climate does not change when the seasons change, but the weather changes with the seasons. The purpose of this activity is to engage youth in thinking of weather vs. climate and related issues and ideas. So "single occurrence" should always be checked when "weather" was checked; and "an average" should be checked when "climate" was checked.

FACILITATING THE ACTIVITY

Help youth identify the difference between weather and climate. They might need help in thinking about climates different than Indiana's climate.

SHARE WHAT HAPPENED

Question: Did you find it difficult to decide in the statements in Table 1 which were about weather and which were about climate?

ANSWER:
Some youth might have difficulty determining where the X's belong, especially if they do not clearly understand that weather is defined over a short time frame, and climate is the average temperatures over 30 years or more.

Question: Are the clothes you wear today determined by the weather or the climate?

ANSWER:
Determined by the weather.

Question: Are the clothes you keep in your closet and dresser determined by the weather or the climate?

ANSWER:
Our clothing choices are based on the weather we expect to have for any season, so they are determined by climate.

Question: Do you have clothes you don't wear very much? If so, why?

ANSWER:
The weather in temperate climates varies a lot over the year, so clothes are needed for hot and cold weather. Sometimes clothes might be needed for very cold temperatures, even though they are not needed very often.

Announcer's Statement	(choose one) WEATHER	CLIMATE	Reason Single Occurrence	An Average
Yesterday the high was 55°F and the low was 43°F.	X		X	
Today we are expecting a high of 61°F.	X		X	
That is 10 degrees above the normal high of 51°F for this date.		X		X
We will have clear skies today with no rain forecast for the next three days.	X		X	
We usually would have 4 inches of rain this month.		X		X
Hurricane season is beginning in the tropics.		X		X
There is a tropical storm developing in the Atlantic Ocean.	X		X	
To the north there is a massive snowstorm developing.	X		X	
We do not normally see a snowstorm like this at this time of year.		X		X
Tornado season is upon us and we should be prepared.		X		X
We usually expect four or five tornado outbreaks to occur this month.		X		X
This morning a tornado damaged a building in the plains.	X		X	

APPLY

Question: Pick another time of year. What clothes would you wear outside at that time of year?

ANSWER:
The clothing items should be appropriate for the time of year.

GENERALIZE TO YOUR LIFE

Question: Would you keep all the same clothes in your closet if you lived in Hawaii? What would be different?

ANSWER:
In Hawaii youth would have more hot-weather clothing and only a few items for cooler weather.

Question: Would you keep all the same clothes in your closet if you lived in Alaska? What would be different?

ANSWER:
In Alaska youth would have more and heavier cold-weather clothing and very little clothing for hot days.

CONNECTIONS

There are many different classification systems for climate. Most climate categories are based on either rainfall/temperature or on plant/animal life. The most popular classification system is the Koppen Climate Classification System.

LIFE SKILLS

- Weather Skills: Determining the difference between weather and climate.
- NGSS: ESS2.D
- Success Indicator: Youth understand that weather is what's happening now, and climate is an average of weather.

WHERE IS THE HEAT?

Big Picture: Many elements like clouds, rain, and the earth's environment affect temperatures, but the driving force is the relationship of the earth to the sun. In the summer the Northern Hemisphere is tilted toward the sun and receives direct, concentrated rays. In the winter the Northern Hemisphere is tilted away from the sun, so the sun's rays are spread over a large area, with the direct concentration hitting the Southern Hemisphere. The tilt of the earth also causes differences in the amount of sunlight received during each season, from over 15 hours of sunlight in the summer, to close to only 9 hours in the winter. The difference in concentration of the sun's rays and seasonal daylight creates differences in temperatures.

Distance from the sun has little to no effect on the heating of the earth. During summer in the Northern Hemisphere the earth is about 3 million miles farther from the sun than in winter. The average distance of the earth from the sun is about 93 million miles, so a 3 million-mile difference is close to insignificant.

FACILITATING THE ACTIVITY

In this activity youth might need your hands for assistance. Youth are capable of completing each section of the activity; however, you might offer to hold the flashlight while they draw or assist with measuring to ensure consistency. Let youth think about how the activity relates to the changing concentration of the incoming sunlight and the time the sun shines during each season. They might also need assistance in calculating daylight times.

SHARE WHAT HAPPENED

Question: What happened to the shape of the light when you moved the flashlight?

ANSWER:
The lighted area became bigger.

Question: How many pennies could you fit in each shape?

ANSWER:
Answers will vary depending on the flashlight used. The angled light should cover a bigger area and therefore fit more pennies.

Question: How will the angle of the light affect the temperature?

ANSWER:
The angled light will not heat as well as the direct light.

WEATHER SKILLS

Question: Which season has the shortest amount of daylight?

ANSWER:
Winter

Question: Which season has the longest amount of daylight?

ANSWER:
Summer

Question: Why would the amount of daylight make a difference in the seasons?

ANSWER:
More daylight gives the sun more time to heat the earth.

Question: Which season had the most direct sunlight?

ANSWER:
Summer

APPLY

Question: What effect do you think the angle of the sun has on temperature?

ANSWER:
The driving force for winter and summer temperatures is the relationship of the earth to the sun. In summer the Northern Hemisphere is tilted toward the sun and receives direct, concentrated rays. In the winter the Northern Hemisphere is tilted away from the sun, so the sun's rays are spread over a larger area, and the direct concentration hits the Southern Hemisphere. The tilt of the earth also causes differences in the amount of sunlight received during each season, from over 15 hours of sunlight in the summer, to close to only 9 hours in the winter. The difference in concentration and seasonal daylight creates the differences in temperatures.

GENERALIZE TO YOUR LIFE

Question: How does the amount and direction of sunlight your area receives affect what you wear in January? In June?

ANSWER:
You need warmer clothes in January because temperatures are colder than in June, when we receive more direct, concentrated sun.

FLY HIGHER

Youth are asked to calculate the area of the shapes they sketched. Use the equation, πr^2, to calculate the area of the circle where r is the radius. Use πab to calculate the area of an oval, where a = the distance from the center to the horizontal edge and b is the distance from the center to the vertical edge.

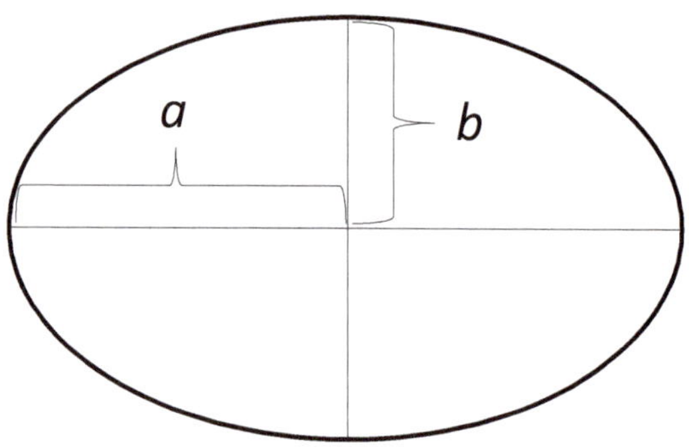

CONNECTIONS

The heat from the sun determines earth temperatures. Any change in this relationship can change how the earth is heated and therefore the dynamics in the atmosphere we call weather.

LIFE SKILLS

- Weather Skills: Understanding that changes in the angle of sunlight and length of daylight contribute to the changing seasons.
- Indiana Science Standards: 5.2.2, 5.2.3
- NGSS: ESS1.B
- Success Indicators: Youth understand how the sun's rays affect the temperature, and how the angle of sunlight and day length causes seasonal changes.

SUPPLEMENTAL INFORMATION AND RESOURCES:

NOAA: www.loc.gov/everyday-mysteries/meteorology-climatology/item/why-is-it-hot-in-summer-and-cold-in-winter/ and https://spaceplace.nasa.gov/seasons/en/.

Note: Sunrise, sunset, and daylight hours data is from NOAA. Use the search box and keywords: sunrise, sunset, and daylight hours.

ACTIVITY FACILITATION AND ANSWERS: LEVEL 2

AIR PRESSURE

Big Picture: Air is an essential part of our lives and all around us, even if we are not aware of its presence. Air is a fluid, and its weight causes air pressure. Differences in air pressure cause movement. In this activity the straw shows how differences in air pressure cause water to move through it. This activity demonstrates the movement of air from areas of high pressure to areas of low pressure. Youth discover that the straw is worthless when no air is in the jar, because there is no difference of air pressure to move water through the straw. With air in the jar, the air expands and pushes water out of the jar when the drinking action lowers air pressure in the straw. The take-home concept here is that differences in air pressure make a drinking straw work. The weather implication is that differences in air pressure cause movement of air in the atmosphere.

FACILITATING THE ACTIVITY

Making the jar apparatus—drilling the metal lid and using the glue gun—requires some adult supervision. For increased safety, put a block of wood under the lid when drilling. The size of the jar does not matter as long as it's glass and the lid and straw seal tightly. The jar should be clean because youth will drink from it.

SHARE WHAT HAPPENED

Test 1

Question: What happened as you tried to use the straw to drink the water?

ANSWER:
Youth should not be able to drink any water.

Question: What happened when you tried to blow bubbles into the jar?

ANSWER:
Youth should not be able to blow air into the jar.

Question: Why do you think this happened?

ANSWER:
Youth might have not have any answers. Good ideas include: the water is taking up that space; there is no air in the jar to push the water out.

Test 2

Question: What happened as you tried to use the straw to drink the water?

ANSWER:
Youth should be able to drink some water but not continue drinking for long.

Question: When you tried to drink the water, were you putting air into or taking air out of the straw?

ANSWER:
Taking air out of the straw.

Question: What happened to the water in the straw immediately after drinking some water?

ANSWER:
After drinking, the water in the straw went back into the jar, and air bubbles went into the jar.

Question: What happened when you tried to blow bubbles into the jar?

ANSWER:
Youth should be able to blow some bubbles into the jar.

Question: When you tried to blow into the straw, were you putting air into or taking air out of the straw?

ANSWER:
Putting air into the straw.

Question: What happened in the straw after you blew bubbles into the jar?

ANSWER:
Water squirted out of the straw.

APPLY

When using a straw, differences in air pressure cause water to move in it. In Part 1 of your experiment no air pressure was in the jar, so there was no difference in air pressure to move the water in the straw in either direction. In Part 2 there was air in the jar, so you could create a difference in air pressure to push the water in the straw.

Question: Which direction does the water move, from low pressure to high pressure or from high pressure to low pressure?

ANSWER:
From high pressure to low pressure

Question: Think about what happened in Part 2 and fill out the following chart:

Question: How can watching a change in air pressure help predict the weather?

ANSWER:
Youth may have some background knowledge that low pressure indicates rainy weather and high pressure indicates clear weather. This information is also in the introduction.

LIFE SKILLS

- Learning to learn, acquiring knowledge critical thinking
- Weather Skills: Air moves from areas of high pressure to areas of low pressure.
- Indiana Standards: 8.2.5
- NGSS: ESS2.D
- Success Indicator: Youth understand that changes in air pressure cause movement.

SUPPLEMENTAL INFORMATION AND RESOURCES

More on aire pressure: https://www.noaa.gov/jetstream/upper-air-charts/common-features-of-constant-pressure-charts.

Action	Pressure in straw: high or low	Pressure in jar: high or low	Movement of water in straw: up the straw or down the straw
Drinking some water from straw	low	high	up the straw
Immediately after drinking some of the water	high	low	down the straw
Blowing air through the straw into the jar	high	low	down the straw
Immediately after blowing air into the jar	low	high	up the straw

GENERALIZE TO YOUR LIFE

Question: Because the weight of air above us causes air pressure, where will air pressure be the greatest: at the base of a mountain or at the top of a mountain?

ANSWER:
Air pressure is greatest at the base of the mountain.

CARBON FOOTPRINTS

Big Picture: Burning fossil fuels is the largest contributor to increasing carbon dioxide. In this activity youth discover their family's estimated contribution to the increase of CO_2. Youth discover their household's day-to-day activities contribute to the carbon dioxide produced by the use of fossil fuels. They also discover actions their family can take to help reduce their impact on this problem.

Many things can be done to reduce a carbon footprint that are not on this website. One is to use reusable canvas bags at the store instead of the store's plastic bags. Plastic bags are not only made from fossil fuels, but producing them also uses energy. Reusing canvas bags saves the energy used to produce plastic bags, and prevents the plastic bags from being burned or thrown into landfills. There are many more ways to help.

FACILITATING THE ACTIVITY

The youth may need some help with this activity. All information can be estimated from utility bills. Make sure the appropriate icons are marked (yearly or weekly, dollars or cubic feet, etc.).

SHARE WHAT HAPPENED

ANSWER:

All questions are interpretive—no right or wrong answers

FLY HIGHER

Youth should find that many things can be done to reduce a carbon footprint beyond what was introduced in this activity. Simple suggestions include reducing or stopping the use of plastic. A reusable shopping bag not made of plastic uses less energy to produce and won't be burned or thrown into landfills. Other strategies include carpooling, riding bikes, walking, insulating the house better, buying local food, using solar panels, reducing wasted food, and composting.

CONNECTIONS

As of 2026 over 40 billion tons of carbon dioxide is added to the atmosphere each year as a result of human activity. This affects not only the atmosphere but also the oceans. Rain reduces the amount of carbon dioxide in the air, as water chemically combines with carbon dioxide to form carbonic acid. The acidity of ocean waters is currently rising, which causes problems for many forms of aquatic life.

LIFE SKILLS

- Learning to learn; wise use of resources
- Weather Skills: Youth recognize their own responsibility in contributing carbon dioxide emissions and that they can help reduce them.
- Indiana Standards: 8.2.6, 8.2.8
- NGSS: ESS2.D, MS-ESS2-5, ESS3.D, MS-ESS3-5
- Success Indicator: Youth want to do their part to reduce carbon dioxide emissions.

SUPPLEMENTAL INFORMATION AND RESOURCES:

EPA website: www.epa.gov/climatechange/ghgemissions/ind-calculator.html

Car fuel economy website: www.fueleconomy.gov/feg/findacar.shtml

CLOUD FORMATION

Big Picture: Clouds form in rising air. For a cloud to form, the air must cool. It cools because air rising into the reduced pressure of the atmosphere expands, and expansion causes cooling.

Our sun is a star that emits shortwave radiation, which we call light energy. Light energy has very little warming effect in the atmosphere, but it warms surfaces of the earth. The warmed surfaces heat the air and cause warmer air to be lifted along frontal boundaries and in thermals. Mountain ranges can also force air to rise; as the air lifts, it expands and cools. If enough moisture is present in the cooling air, water vapor condenses to form a cloud.

FACILITATING THE ACTIVITY

The dry bottle must be completely dry, which might take a day of upside-down drying. The nature of any air/water interaction is that in any closed container of air and water, the air becomes saturated with water vapor. The maximum amount of water vapor exists in that air, and the maximum amount of air dissolves into the water.

Youth will probably not get a very good "cloud" the first time they try this activity, but they'll likely do better on subsequent tries.

Question: How many pumps were necessary to get a cloud that flows out of the bottle?

ANSWER:
Depending on the brand of fizz keeper, this could be as many as 200 pumps.

Question: Why is it important to shake the water in the bottle and let the bottle sit?

ANSWER:
Shaking encourages evaporation to get to saturation.

Question: What is changing in the bottle when air is being pumped into it?

ANSWER:
The air pressure in the bottle is increasing.

Question: What kind of results did you get with the dry bottle?

ANSWER:
Little to no condensation or cloud.

Question: Why is there a difference between the dry bottle and the wet bottle?

ANSWER:
There is not enough water vapor in the dry bottle to get a cloud.

Question: Why can birds soar and even climb higher without flapping their wings?

ANSWER:
They are caught in a thermal, and the rising air allows them to soar.

Figure 3. Thermal

Question: What is causing the air to lift?

ANSWER:
The warming, less dense air rises in the cooler air. This is caused by the difference in air density.

Question: What is happening to the temperature of that rising air?

ANSWER:
As the air rises, it cools.

Figure 4. Mountain Effect

Question: What is causing the air to lift?

ANSWER:
The air is forced to rise up the slope of the mountain range.

Question: What is happening to the temperature of that rising air?

ANSWER:
As the air rises, it cools.

Figure 5. Cold Front

Question: What is causing the air to lift?

ANSWER:
The colder air pushes the warmer, less dense air up. This is caused by the difference in air density.

Question: What is happening to the temperature of that rising air?

ANSWER:
As the air rises, it cools.

Figure 6. Warm Front

Question: What is causing the air to lift?

ANSWER:
The warmer, less dense air is being pushed over the colder air. This is caused by the difference in air density.

Question: What is happening to the temperature of that rising air?

ANSWER:
As the air rises, it cools.

APPLY

Question: How does the bottle and Fizz-Keeper Pump activity relate to the figures?

ANSWER:
Youth should see a cloud form in the bottle due to the increased pressure and start to understand why clouds form.

GENERALIZE TO YOUR LIFE

Question: Why do glider pilots look for small puffy clouds on a sunny day?

ANSWER:
They know that if the clouds are there, the air under them must be rising.

CONNECTIONS

As air rises over mountain ranges, clouds and rain form on that side (windward) of the range. Air sinks on the other side, and as air sinks, it warms up. This creates air that is warmer and dryer. The U.S. western mountain ranges create rain shadow zones to their east. This is why we have vast desert areas east of these mountain ranges, but rainfall is common to their west. The forests in western Washington are rain forests.

LIFE SKILLS

- Critical thinking; problem solving
- Weather Skills: Clouds form in air that is lifted. This lifting causes the air to expand, which causes cooling. There are three basic ways that lifting can occur: thermals, fronts, and mountain ranges.
- Indiana Standards: 8.2.3.
- NGSS: ESS2.D
- Success Indicator: Youth can explain how clouds form.

CLOUD TYPES

Big Picture: The main goal of this activity is to make youth aware of the sky and changes that occur there. They become sky watchers. Most clouds in Indiana form because air is lifted, and this lifting provides cooling. Lifted air expands in the reduced air pressure aloft. This expansion causes cooling of the air (the Ideal Gas Law). Multiple layers of clouds indicate multiple layers of air being lifted.

FACILITATING THE ACTIVITY

Cloud type identification can be difficult, because clouds do not always conform to the classification system. While most clouds have characteristics that place them into one of the 10 basic types, others seem of two different types, and some do not seem to fit at all. Youth should know that their best guess is acceptable. An example might be a sky with small cumulus-type clouds that seem to be layered, but only loosely. One might call those clouds cumulus, and another might call them stratocumulus—and that's fine.

SHARE WHAT HAPPENED

ANSWER:
Interpretive—no right or wrong answers

APPLY

Question: In which example will you find cumulonimbus clouds?

ANSWER:
Cumulonimbus clouds are found in the cold front.

Question: In which example will you find fair-weather cumulus clouds?

ANSWER:
Cumulus clouds are formed by uneven heating.

Question: In which example will you find nimbostratus clouds?

ANSWER:
Nimbostratus clouds are formed by a warm front.

GENERALIZE TO YOUR LIFE

Question: Dark, thick stratus clouds

ANSWER:
Rain or snow

Question: Small, puffy cumulus clouds

ANSWER:
Sunny, fair weather

Question: Towering dark cumulonimbus clouds

ANSWER:
Heavy wind and rain, possible thunderstorms

Question: High wispy cirrus clouds

ANSWER:
Fair weather

LIFE SKILLS

- Learning to learn; keeping records
- Weather Skills: Clouds appear in different forms in the sky that are categorized into types, with some understanding how the different types form.
- Indiana Standards: 8.2.3.
- NGSS: ESS2.D
- Success Indicator: Youth recognize cloud types in the sky.

SUPPLEMENTAL INFORMATION AND RESOURCES

Cloud Spotter Wheel (PDF): www.noaa.gov/jetstream/global/jet-stream

Cloud Chart: www.noaa.gov/jetstream/clouds/nws-cloud-chart

The abbreviations below are often used for the 10 most common clouds.

Cloud Type	Abbreviation
Nimbostratus	Ns
Cirrocumulus	Cc
Cumulus	Cu
Cirrus	Ci
Cirrostratus	Cs
Cumulonimbus	Cb
Altocumulus	Ac
Stratocumulus	Sc
Altostratus	As
Stratus	St

COLD FRONTS

Big Picture: This activity shows what cold fronts are and how they function. Cold water, coming from the west (left), approaches warmer air ahead of it to the east (right). In the jar, as the denser cold water approaches, it provides a lifting mechanism that pushes the warmer water above it. This lifting mechanism acts like a wedge, pushing the warm water ahead of it upward. This is important for precipitation to develop; air needs to be lifted, which forces it to cool as it expands in the lower pressure of the higher altitude. (The air's expansion cools its temperature.) This creates an unstable situation that results in the warm air continuing to lift until it is colder than its surroundings. This allows the condensation of water vapor, which forms clouds, precipitation, and thunderstorms.

This air might never have lifted on its own unless the cold front gave it a push. These lifts occur along the entire front, which is why showers and thunderstorms tend to form along most of the frontal boundary, usually forming a narrow band of clouds and storms.

Cold fronts form to the south of a low pressure system because its circulation (cyclonic or counterclockwise) moves cold air from the north down behind the developing front, and moves warm air up from the south ahead of the front. The increasing differential in temperatures on either side of the low creates a boundary that forms into a cold front. Winds also come from different directions on both sides of the cold front. Winds typically blow from the south on the warm side and from the west on the cold side, mainly due to the winds from the approaching surface low.

FACILITATING THE ACTIVITY

This activity can be messy, so do it over a pan. Make sure the two jars have same-sized openings; otherwise, water will leak out. The most difficult part comes when the barrier between the two jars is removed. Take care that the seal between the jars remains tight by holding them together with your hands. Some leakage is expected, and the index card might begin to fall apart. Losing some water is normal. The jars need to be backlit by a window or light to see the results better.

SHARE WHAT HAPPENED

ANSWER:
Interpretive—no right or wrong answers

APPLY

Question: How similar is the diagram of a cold front, in Figure 5, to what you observed with the two colored waters in your jars?

ANSWER:
Youth should observe the same type of initial movement. The blue water fills the bottom, but the back-and-forth movement would not occur in the atmosphere.

Question: Why are clouds forming at the boundary between the two air masses?

ANSWER:
The warm air is being lifted. That forces the air to cool, causing water vapor to condense into clouds.

Question: In the satellite picture of a cold front approaching Indiana, why is the band of clouds narrow?

ANSWER:
This occurs because of the rapid lifting of the warm air ahead of the front.

GENERALIZE TO YOUR LIFE

Question: You see an approaching cold front on a weather map. What kind of weather changes are you most likely to observe when the front arrives?

ANSWER:
Youth should expect cloudy weather with showers and thunderstorms.

Question: Why?

ANSWER:
This happens because of the rapid lifting of warm air ahead of the front.

CONNECTIONS

A rare weather setup allowed one of the strongest cold fronts in history to form on November 11, 1911 (11-11-11, similar to the triple date experienced in 2011). On the warm side, several strong tornadoes formed in the lower Great Lakes. After the front passed, some of the exact same locations experienced blizzard conditions! The front was so dramatic that Oklahoma City set a record high and a record low on the same day, when the temperature fell from 83° to 19°, a change of 64 degrees.

LIFE SKILLS

- Acquiring knowledge, critical thinking, reasoning
- Weather Skills: Understand the formation and nature of a cold front.
- Indiana Science Standards: 8.2.5
- NGSS: ESS2.D
- Success Indicators: Youth understand that the cold water pushes the warm water over it because the cold water is denser. Cold fronts work in a similar fashion; cold air from the west pushes warm, less dense air above it, providing buoyancy. With an approaching cold front, showers and thunderstorms are likely, along with a shift in wind direction and dropping temperatures.

EARTH'S ROTATION

Big Picture: The Coriolis effect, which has major effects on weather and wind patterns, is sometimes referred to as a "force." The Coriolis effect can be treated mathematically as a force, but no actual forces cause it. The apparent change in motion is caused by earth's surface moving at different speeds—fastest at the equator and gradually slower to its north or south. The only place on earth you can travel with no Coriolis effect is straight east or west along the equator. You cannot travel "straight" east or west along any other latitude line. To demonstrate, place a string around the equator of a globe. It remains straight. Now place the string along another latitude on the globe. It is no longer straight, due to the curve of the earth. It remains straight only at the equator.

Wind is deflected to the right north of the equator and to the left south of the equator because of the Coriolis effect. This is why tropical storms cannot cross the equator. The earth's rotation is counterclockwise in the Northern Hemisphere and clockwise in the Southern Hemisphere.

FACILITATING THE ACTIVITY

Rotation—Youth are instructed to turn the globe counterclockwise because the Northern Hemisphere has a counterclockwise rotation.

Timing—The earth rotates once in 24 hours. Watching the hour hand of the clock, which rotates twice as fast as the earth, illustrates how slowly the earth turns. It is hard to imagine a clock hand 4,000 miles long. If the hour hand was 4,000 miles long (approximate radius of the earth), its tip would be moving at 2,000 mph!

Position—Youth need at least two other people to play catch while they turn the roundabout and retrieve the errant ball. This is a classic demonstration of the Coriolis effect. The youth should discover that the ball appears to arc to the right when tossed as the roundabout rotates counterclockwise (Northern Hemisphere) and to the left when rotating clockwise (Southern Hemisphere). Air moving across the surface of the earth (wind) is like the ball being tossed across the surface of the roundabout.

WORKSHEET QUESTIONS

Rotation 1

Question: What do you notice about the two arrows?

ANSWER:
They point in opposite directions.

Question: What is the difference in rotation between the Northern and Southern hemispheres?

ANSWER:
One moves clockwise and the other rotates counterclockwise.

Question: Did you see the minute hand move while you were watching it?

ANSWER:
Youth should be able to see the minute hand move while watching it for one minute.

Question: Did you see the hour hand move while you were watching it?

ANSWER:
Youth probably won't see the hour hand move while watching it for one minute. They would need to watch it for much longer to observe its movement. This activity is intended to give them an idea of how scale affects our perception.

Question: The hour hand on a clock rotates once

Location (Latitude)	Distance traveled (24 hours)	Speed (miles/hour)
Equator (0°)	About 25,000 miles	1,042
Indianapolis, IN (about 40°)	About 18,000 miles	750
Anchorage, AK (about 60°)	About 12,000 miles	500

every 12 hours, and it takes the earth 24 hours to rotate once.

ANSWER:
So the hour hand rotates twice as fast as the earth.

Question: What part of the earth travels the fastest? Equator, Indianapolis, Anchorage

ANSWER:
Equator

Question: Is Chicago rotating faster or slower (for someone in Indiana) than you?

ANSWER:
A city to the north is rotating more slowly, because it has less distance to travel in 24 hours. So if Chicago is north of the youth, it is moving slower than the youth.

Question: Is New Orleans rotating faster or slower (for someone in Indiana) than you?

ANSWER:
A city to the south is rotating faster because it has more distance to travel in 24 hours. So if New Orleans is south of the youth, it is moving faster than the youth.

Rotation 2

Question: What happened when you started rotating the roundabout counterclockwise (similar to the motion of the Northern Hemisphere)?

ANSWER:
The youth's friends probably had more difficulty catching the ball and might have missed it a number of times before adjusting to the movement of the roundabout. The ball appears to curve instead of traveling straight. Air moving across the surface of the earth (wind) is like the ball being tossed across the surface of the roundabout.

Question: Did the ball seem to curve?

ANSWER:
Yes

Question: Which way?

ANSWER:
Right. This is because wind is deflected to the right in the Northern Hemisphere.

Question: Now have your friend turn the roundabout slowly in a clockwise rotation. This is like the rotation of the Southern Hemisphere. Note how the ball appears to curve (right or left).

ANSWER:
The ball should appear to curve left.

Question: How do you think the earth's rotational speed affects wind and weather patterns?

ANSWER:
The different rotational speeds affect wind and wind patterns. Watching the weather for the entire United States is a good way to begin understanding how weather patterns move across the U.S., and the relationship of the earth's rotation and wind direction.

SHARE WHAT HAPPENED

ANSWER:
Interpretive—no right or wrong answers

APPLY

Question: Have you ever walked between surfaces that are moving at different speeds, for example, stepping onto a moving sidewalk at an airport or amusement park ride? How does that affect your balance?

ANSWER:
Surfaces that are moving at different speeds require us to adjust our steps. Youth might say they feel unsteady and tend to lose their balance.

Question: How is wind affected in the Northern Hemisphere, which is rotating counterclockwise? Is the wind being deflected to the right or left?

ANSWER:
Wind is deflected to the right north of the equator because of the Coriolis effect. Wind is deflected to the left south of the equator.

GENERALIZE TO YOUR LIFE

Question: What objects might be affected by the Coriolis effect?

ANSWER:
Youth may give a variety of answers: airplanes, cars, balls, water, bullets, runners, toilets, or ocean-going ships.

FLY HIGHER

Question: Explore other things that move across the earth's surface and might be affected by the Coriolis effect, such as airplanes and missiles. The Internet may be your best resource. Youth might need help getting started with their Internet search.

ANSWER:
Common answers include airplanes, toilets, and baseballs. Youth might also find water, bullets, a runner, or ocean-going ships.

CONNECTIONS

The Coriolis effect accounts for most air movement. Friction against the surface of the earth diminishes the Coriolis effect. As altitude increases, the Coriolis effect is often the only force acting on the air's movement.

Some sources cite the first practical observation of the Coriolis effect as long-range cannons that did not respond to the known physics of cannon flight, missing targets to the right. This could be true if no other factors were affecting their flight; however, a crosswind would have greater effect on a cannon ball. The Coriolis effect does not deflect cars, because they are in constant contact with the surface, and friction negates the effect. Airplanes must adjust to the Coriolis effect, although more so to wind patterns in the atmosphere. The earth moves too slowly for the Coriolis effect to have much influence on anything that travels a short distance and is affected by other factors. Baseballs, bullets, and other short-lived motions are more influenced by other factors.

LIFE SKILLS

- Learning to learn, acquiring knowledge
- Weather Skills: The Coriolis effect deflects wind to the right in the Northern Hemisphere and to the left in the Southern Hemisphere.
- Indiana Standards: 8.2.1, 8.2.9
- NGSS: ESS2.D
- Success Indicator: Youth understand that the earth's rotation causes the Coriolis effect, which causes wind to deflect to the right in the Northern Hemisphere and to the left in the Southern Hemisphere.

SUPPLEMENTAL INFORMATION AND RESOURCES

Gaspard Coriolis: www.britannica.com/EBchecked/topic/137641/Gustave-Gaspard-Coriolis

Calculating rotational speed: http://imagine.gsfc.nasa.gov/docs/ask_astro/answers/970401c.html

GLOBAL WINDS

Big Picture: The energy from the sun and the earth's rotation (Coriolis effect) combine to form global wind patterns. Each wind belt exists because of these two factors, which affect the climate of every area on the earth. Within the trade winds, hurricanes develop. In the prevailing westerlies, cyclonic storms move from west to east in waves, creating fronts and storms. Along the equator, the area known as the doldrums, warm, moist air rises to create convective storms. Along the horse latitudes, constantly sinking air creates the world's major deserts. These belts are not static. They are influenced by the seasons, landmasses, and ocean currents to constantly move their boundaries.

FACILITATING THE ACTIVITY

If youth are struggling with how the earth's rotation influences wind movement, recommend they complete or review the Level 2 Earth's Rotation activity.

WORKSHEET QUESTIONS

Youth are not expected to know these answers, but to explore how global wind patterns affect weather.

Question: How might the sun have an effect on global wind patterns?

Question: How might the rotation of the earth affect global wind patterns?

ANSWER:
The first two questions are interpretive—intended to encourage youth to form an opinion before they begin the activity. There are no wrong answers.

Question: Why is the air less dense at the surfaces near the equator than those at the North and South Poles?

ANSWER:
The sun warms the air at the equator more than at the poles, and air at the equator becomes more humid. Warm, humid air is less dense than cool, dry air.

Question: Why is warm, moist air rising above the surfaces near the equator?

ANSWER:
The air rises because it is less dense than the air away from the equator.

Question: In what direction is the surface air shown in Figure 2 moving—toward areas near the equator or those away from the equator?

ANSWER:
The surface air moves toward areas near the equator.

Question: In what direction is the air high above the surface moving – toward areas near the equator or those away from the equator?

ANSWER:
The upper air moves away from from areas near the equator.

Question: What prevents simple global surface winds from blowing from the North and South Poles to the equator?

ANSWER:
Earth's rotation prevents these winds from traveling in a straight line.

Question: Describe how air moves in the Hadley cell region.

ANSWER:
Air rises at the equator and sinks at 30° N and S latitude.

Question: Describe how air moves in the Ferrel cell region.

ANSWER:
Air sinks at 30° latitude and rises at 60° latitude.

Question: Describe how air moves in the polar cell region.

ANSWER:
Air rises at 60° latitude and sinks at the poles.

Question: Why is the area along the equator known as the doldrums?

ANSWER:
This area of rising air has very little surface winds, leaving sailing ships floating aimlessly.

Question: Why do deserts form along 30° latitudes?

ANSWER:
Air generally sinks along 30° latitude and clouds do not form in sinking air, so very little rainfall occurs along this band.

Convection Cell	Within Latitudes:	Name of Surface winds:	Direction of winds
Hadley	0°- 30°	Trade Winds	East to west
Ferrel	30° - 60°	Prevailing Westerlies	West to East
Polar	60° - 90°	Polar Easterlies	East to west

Question: Indicate the locations of the following in the following figure:

ANSWER:

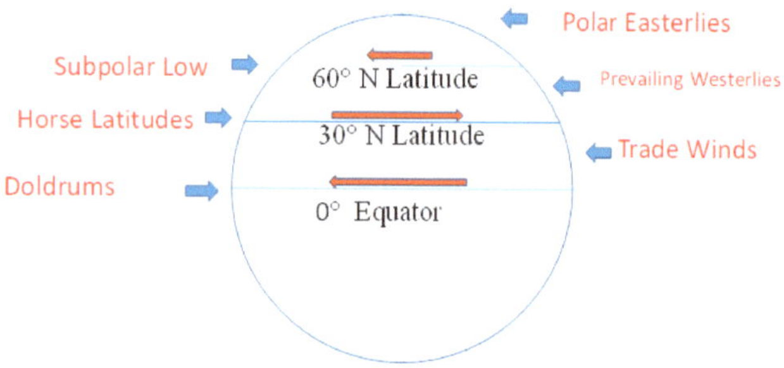

CONNECTIONS

Christopher Columbus traveled down the coast of Africa to the Canary Islands in 1492 to avoid possible problems with the Portuguese. (The Portuguese frowned on other nationalities sailing out into the Atlantic Ocean, and Columbus's expedition was financed by Spain.) This put Columbus in an area mostly absent of winds, and he waited several days to begin his journey. Once he started, he did not realize that a compass does not always point exactly toward real north, a mistake that took him deeper into the trade winds. This resulted in winds constantly pushing his ships to the west—what he wanted, but worrisome to his sailors. They never thought they might fall off the earth; it is a social myth that Columbus was trying to prove the earth was round. They knew in 1492 that the earth was round. The sailors really worried that the wind never changed directions, so how would they ever return? It turns out that Columbus was lucky to make his journey in the trade winds. The Portuguese had already tried sailing west but were unsuccessful because their attempts were to the west of Portugal, in the prevailing westerlies, where winds pushed them back to the east.

SHARE WHAT HAPPENED

Question: What force causes air to move?

ANSWER:
The sun

APPLY

Question: Name a city, town, or county that usually gets the same weather you observe in your home's area after you experience it.

ANSWER:
This question is intended to have youth think about where the prevailing winds at their home come from. Watching national weather patterns is the easiest way to observe this. Prevailing winds in the Midwest are generally from the northwest.

GENERALIZE TO YOUR LIFE

Question: How can understanding global wind patterns help you prepare for the next few days' weather?

ANSWER:
Storms and daily weather can be tracked because they generally follow the same patterns. Meteorologists understand these patterns and how other influences can affect the weather when they make predictions.

LIFE SKILLS

- Acquiring knowledge
- Weather Skills: Describe the global wind patterns, and explain how these patterns are a result of the sun's energy and the earth's rotation.
- Indiana Standards: 8.2.1
- NGSS: ESS2
- Success Indicator: Youth understand global wind patterns and the types of climate that result.

GREENHOUSE EFFECT

Big Picture: The greenhouse effect prevents infrared radiation from exiting into space, which maintains the equilibrium temperature of the planet. Certain gases called greenhouse gases—CO_2, water vapor, methane, and nitrous oxide—have chemical properties that allow them to absorb and re-emit incoming infrared radiation. So these gases reabsorb infrared radiation from the earth's surface and re-emit it back toward the surface. Having an excess of these chemicals in the atmosphere causes too much of the infrared radiation to stay, increasing temperatures beyond where they should be, and at too quickly a rate, for many plants and animals to adapt.

Concentrations of carbon dioxide in the chart are in parts per million. Currently the concentration of carbon dioxide in the atmosphere is nearing 400 ppm. It varies across the earth, being greater in areas of greater populations. Concentrations of 400 ppm only represent .04 percent of the atmosphere. Water vapor is also a greenhouse gas, but concentrations of water vapor continually change, and it has other warming and cooling effects on the atmosphere. Not considering water vapor, the total concentration of greenhouse gases in the atmosphere is less than half of 1 percent. More than 99% of the atmosphere is made up of gases that do not absorb heat energy, allowing it to escape into space. That tiny fraction of greenhouse gases is responsible for keeping our planet at its present temperature equilibrium. That is why small increases of greenhouse gases can potentially cause great changes.

FACILITATING THE ACTIVITY

The containers need to be identical and large enough to hold the thermometer. The predictions are intended to encourage youth to estimate temperatures and consider what they will find during this activity.

SHARE WHAT HAPPENED

Question: Were your predictions correct?

ANSWER:
Interpretive—no right or wrong answer

Question: Which container had a greater temperature after the first 30 minutes?

ANSWER:
Container A, the one with the plastic wrap.

Question: What caused this difference?

ANSWER:
The plastic wrap kept the heat from escaping.

Question: After the second 30-minute period, which container had the greater temperature?

ANSWER:
Container A still has the higher temperature.

Question: What caused this difference?

ANSWER:
The plastic wrap kept the heat from escaping.

Question: How is the plastic wrap like greenhouse gases?

ANSWER:
Both keep the heat energy from escaping, the container or earth's atmosphere.

APPLY

Question: How will an increase of greenhouse gases affect the temperature of the atmosphere?

ANSWER:
Increased greenhouse gases retain more heat. This causes the average global temperature to increase.

CONNECTIONS

The current concentration of the primary greenhouse gas, carbon dioxide, is approximately 395 ppm (parts per million). Most atmospheric scientists agree this concentration needs to be reduced to 350 ppm to avoid damaging and costly side effects of global warming.

LIFE SKILLS

- Problem solving, critical thinking, acquiring knowledge
- Weather Skills: Use thermometers to record data and analyze that data using graphs.
- Indiana Standards: 8.2.6, 8.2.8
- NGSS: ESS2.D, MS-ESS2-5, ESS3.D, MS-ESS3-5

- Success Indicator: Youth recognize the importance of greenhouse gases in the atmosphere and how their increase could affect the temperature of the atmosphere.

SUPPLEMENTAL INFORMATION AND RESOURCES

Image of a greenhouse: http://climatekids.nasa.gov/greenhouse-effect/

Mauna Loa Data: http://www.esrl.noaa.gov/gmd/ccgg/trends/

HUMIDITY

Big Picture: The idea of latent heat can be confusing. We expect that when heat energy is added to matter, the temperature of the matter should rise. That is true as long as the matter does not change states. However, when matter changes state, temperature does not change, even though energy is added or removed. This absorption or release of energy is latent heat. Evaporation has a cooling effect as heat energy is absorbed, and condensation has a warming effect as heat energy is released. When water is boiling, the liquid water stays at the boiling point even if more heat energy is added. (Turning up the heat on the stove only causes the water to boil more quickly.) The resulting water vapor is also at the boiling point. Heat energy causes the temperature of the water vapor to rise. This can be more easily seen in a change from ice to liquid. As ice melts in a glass, the ice water mixture stays at or near the freezing point until most of the ice is melted.

When using a hygrometer, the difference between the dry bulb and wet bulb gives a precise idea of how humid the air is. In humid air, the difference is slight. As the air becomes dryer, evaporation quickens, and the difference between the dry bulb and wet bulb increases.

FACILITATING THE ACTIVITY

Youth need to be careful when boiling the water. Make sure the thermometer stays away from the bottom and sides of the pan to measure only the water temperature.

Question: What energy is used to raise the temperature of the water?

ANSWER:
The energy used depends on the type of stove used (electric or gas).

Question: What happened to the temperature readings when the water started boiling?

ANSWER:
The temperature stays at 212°F, the boiling point of water.

Question: Why?

ANSWER:
At 212°F water begins to boil—go through a phase change to become a gas (steam). This phase change requires energy. When matter changes state, temperature does not change, even though energy is added or removed.

SHARE WHAT HAPPENED

Part 1:

Question: Was the wet-bulb or dry-bulb temperature higher?

ANSWER:
Thermometer 1, the dry bulb thermometer, should record the higher temperature.

Part 2:

Question: What happened to the temperature once the water began to boil?

ANSWER:
The temperature is constant when the water is boiling. It does not continue to rise.

APPLY

Question: Why do you think the wet-bulb temperature is generally lower than the dry-bulb temperature?

ANSWER:
Evaporation is occurring on the wet bulb. Since evaporation has a cooling effect, the wet-bulb temperature is lower.

Question: Will water evaporate slower or more quickly when air is very dry?

ANSWER:
When the air is drier, water evaporates more quickly.

Question: When your body begins to overheat, you perspire. Why does perspiration help cool your body?

ANSWER:
Sweat, water that moves to the surface of your skin, evaporates and removes heat from your body. This is your body's natural cooling system.

Question: What happens to perspiration on a hot, humid day?

ANSWER:
On a hot, humid day perspiration does not evaporate as well as on a dry day, so it is not removed as fast. This results in slower cooling.

Question: In desert towns, people often talk about the weather having a "dry heat." Why would low humidity make hot weather more comfortable?

ANSWER:
Perspiration can more easily evaporate off your skin in low humidity than in high humidity. This helps you cool.

Question: Why do people complain, "It's not the heat, it's the humidity?"

ANSWER:
In humid weather you are uncomfortable because perspiration, the body's natural cooling mechanism, is slowed.

GENERALIZE TO YOUR LIFE

Question: When you go swimming on a hot day, your body adjusts to the cooler temperature of the water, and then the water feels fine. Why do you feel colder when you get out of the water and into the hot air?

ANSWER:
When you get out of the water, you are wet, and the evaporating water cools your body.

Question: If you are cooking something in boiling water, can you get the water hotter by increasing the burner's heat setting?

ANSWER:
No. Once the water is boiling, the temperature cannot rise above 212°F (100°C). It stays at the boiling point until it is evaporated (becomes a gas).

FLY HIGHER

Question: Did the temperature change as the ice melted?

ANSWER:
The temperature should not change, or will change very little, as the ice melts.

CONNECTIONS

Water boils at different temperatures depending on the air pressure. Water boils at about 200°F in Denver, which has an elevation of one mile. Things cook more slowly there. If you want to cook something at a higher temperature, you can add salt to the water, which raises the boiling point a few degrees. If you need to cook something in water at much higher temperatures, you will need a pressure cooker. The higher pressure increases the boiling point of water.

LIFE SKILLS

- Problem solving, critical thinking, acquiring knowledge
- Weather Skills: Evaporation has a cooling effect, and condensation has a warming effect. This absorption or release of energy is called latent heat.
- Indiana Standards: 8.2.2 8.2.4
- NGSS: ESS2.C, ESS2.D
- Success Indicator: Youth understand that heat energy is absorbed during evaporation and released during condensation.

HURRICANES

Big Picture: Hurricanes are violent storms that form over the Atlantic Ocean. They are also called typhoons or cyclones, depending on where they occur. The scientific term for all of these storms is tropical cyclone, but only tropical cyclones that form over the Atlantic Ocean or eastern Pacific Ocean are called hurricanes. Whatever they are called, tropical cyclones all form the same way: only over ocean water near the equator, using the warm, moist air as fuel. The warm air rises, causing an area of lower air pressure below. Air with higher air pressure pushes in from surrounding areas to the low-pressure area. That new air becomes warm and moist and rises, too. As the warm air continues to rise, surrounding air is pulled in to take its place. As the warmed, moist air rises, it cools and forms clouds. The whole system of clouds and wind spins and grows, fed by the ocean's heat and water evaporating from the surface.

Because of earth's rotation on its axis, storms that form north of the equator spin counterclockwise, and storms that form south of the equator spin clockwise. As a storm system rotates faster and faster, an eye forms in the center. It is calm and clear in the eye, with very low air pressure. Higher pressure air from above flows down into the eye of the storm.

This activity is designed to give youth a greater understanding of hurricanes and practice plotting latitude/longitude coordinates on a map. They track the evolution of hurricanes and connect what causes them to intensify and weaken. Hurricanes develop from low pressure systems that the trade winds push westward over tropical ocean waters. More heat energy and moisture is added to the system, which progresses into a tropical depression, a tropical storm, and then into a hurricane. Most systems curve to the right (northward) because of the rotation of the earth (Coriolis effect), and then turn toward the east as they enter the eastward flow of the mid-latitude prevailing westerlies. This is not always the case, as Hurricane Sandy showed. Youth should understand that as pressure decreases, the intensity of a hurricane increases. Finally, they should realize that making landfall significantly decreases a hurricane's intensity, because it is no longer over the warm water that is its source of essential moist air.

FACILITATING THE ACTIVITY

Make sure the youth plot the coordinates correctly and do not get latitude and longitude backwards. Hurricane tracks should look similar to the one that follows. This chart was made using the NOAA Atlantic Basin Hurricane Tracking Chart.

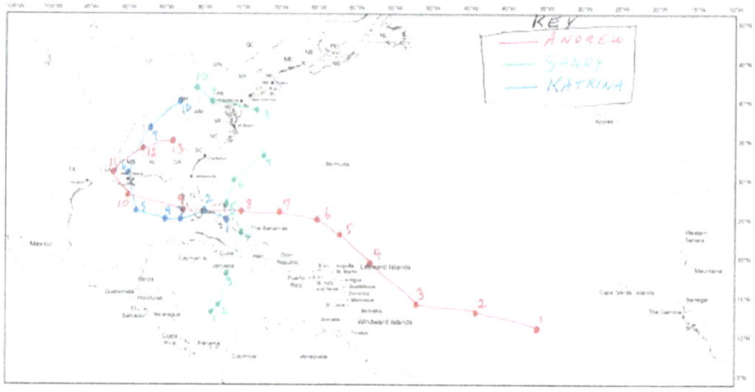

This activity was adapted from National Oceanic and Atmospheric Administration/Department of Commerce resources shown. See these resources for more information.

- https://www.weather.gov/ilm/HurricaneFlorence
- Atlantic Tracking plot: http://www.nhc.noaa.gov/pdf/tracking_chart_atlantic.pdf
- Hurricane in history: https://www.nhc.noaa.gov/outreach/history

SHARE WHAT HAPPENED

Question: What happens to the wind speeds as the air pressure falls?

ANSWER:
Wind speeds increase.

Question: What kind of surface was each hurricane over when it reached its highest wind speeds?

ANSWER:
Over the ocean.

Question: What category did each hurricane reach?

ANSWER:
Sandy, 3; Katrina, 5; Andrew, 5

Question: What happened to each storm's intensity as it made landfall? Why do you think this occurred?

ANSWER:
The intensity decreased rapidly because of the absence of warm ocean waters.

Question: What is the major difference between the paths of Hurricane Sandy and the other two hurricanes?

ANSWER:
Hurricane Sandy traveled north and then back west. The other two traveled west, turned north, and then turned east.

Question: How do you think hurricanes develop when they typically start as strong storms?

ANSWER:
Hurricanes form over ocean water near the equator and use the warm, moist air as fuel. The warm air rises, causing an area of lower air pressure below. Air from surrounding areas with higher air pressure pushes in to the low-pressure area. That new air becomes warm and moist and rises, too. As the warm air continues to rise, the surrounding air is pulled in to take its place. As the warmed, moist air rises, it cools and forms clouds. The whole system of clouds and wind spins and grows, fed by the ocean's heat and water evaporating from the surface.

APPLY

Question: What two continents do hurricanes strike the most?

ANSWER:
North America and Asia.

Question: For the most part, at what latitude do hurricanes appear to start?

ANSWER:
Note: Compare this to your tracking map. Between the latitudes 10 and 20.

Question: If you want to avoid hurricanes, should you live on the East Coast or the West Coast?

ANSWER:
The West Coast has fewer hurricanes than the East Coast.

GENERALIZE TO YOUR LIFE

Question: Do you know anyone who lives along the Gulf of Mexico or on the East Coast? What should be part of a plan to prepare for a hurricane in these areas?

ANSWER:
Plans could include having an emergency kit; bringing loose items like outdoor furniture indoors; boarding up windows; knowing evacuation routes; protecting valuables and important documents.

CONNECTIONS

Surprisingly few movies feature hurricanes. One that covers the aftermath and problems with Katrina well is Spike Lee's *When The Levees Broke*. It is rather long and broken into four parts, not all of which need to be shown. The film is highly acclaimed, however, and shows how devastating hurricanes can be.

NOAA, The U.S. National Oceanic and Atmospheric Administration, has excellent information about weather safety:

Weather Safety webpage, www.weather.gov/safety

Hurricane Preparedness webpage, www.nhc.noaa.gov/prepare/

LIFE SKILLS

- Critical thinking, evaluating data, processing information.
- Weather Skills: Track hurricane paths on a map using latitude and longitude data and gain an understanding of the nature of hurricanes by analyzing the data.
- Indiana Standards: 8.2.3

- NGSS: ESS2.D
- Success Indicators: Youth can correctly plot the hurricanes and discover that landfall weakens them. They also see the connection between pressure and wind speed. They understand that hurricanes tend to strengthen over warmer waters and weaken over land.

MAKING WEATHER INSTRUMENTS

Big Picture: Youth build their own weather instrument(s): a rain gauge, wind vane, barometer, hygrometer, and anemometer. They should be encouraged to use the instruments they make to collect weather data (see the Using Weather Instruments activity). These instruments can be purchased, but youth will understand the weather better and appreciate the accomplishment of making their own. And while wind speed can be estimated using the Beaufort Scale, and the direction of the wind can be observed by noting the direction it blows smoke, small twigs, etc., making observations using wind-measuring devices is more scientific.

FACILITATING THE ACTIVITY

Rain Gauge: Youth are directed to pour the contents of their rain gauge into a smaller diameter can (or cans). This increases the accuracy of their measurement because it magnifies the depth of precipitation to be measured. The precipitation can be calculated by multiplying the measured depth times a factor that shows the relationship between the inner diameter of the collecting funnel and the measuring can. (Note: If the smaller can is not used, the depth of precipitation is the same as the measured amount because D = d, so P = M.) You can help youth understand how using the smaller cans gives a more accurate measurement by comparing the depth of water in the large can and calculating the depth by measuring the depth in the smaller can, and calibrating it with the given formula.

Wind Vane: Finding a good place to use the wind vane can be difficult. Trees and buildings are obstructions that change wind speed, so the reading won't represent the actual wind speed.

Anemometer: Explain that the anemometer measures wind speed like a car's odometer indicates the speed of a car.

Barometer: This type of barometer is sensitive to temperature changes, so keep it in an area that has a steady temperature. A barometer should stay indoors; air pressure is the same inside as it is outside. Youth might be surprised to see that barometer readings change very little. However, small changes are important and can indicate that severe weather is on the way.

Hygrometer: A hygrometer is used to measure the moisture content in the air. Scientists often use a sling psychrometer, which reaches the wet-bulb and dry-bulb temperatures faster than the hygrometer youth make in this activity. These temperatures are used to determine the relative humidity and dew-point temperatures.

SHARE AND APPLY QUESTIONS

ANSWER:
Interpretive—no right or wrong answers

GENERALIZE TO YOUR LIFE

Question: Why are people so interested in the weather?

ANSWER:
People are interested in the weather for many reasons: to decide what to wear, if they need an umbrella, or if an outdoor activity should be canceled. Knowing that severe weather is approaching and how to navigate the weather can be lifesaving. Severe weather includes extreme heat, snowstorms, tornadoes, hurricanes, drought, and other conditions. Media (print, radio, and Internet) provide weather information to help people make daily decisions. When severe weather occurs, this information is amplified to help keep people safe—and because increasing weather concerns raises our interest in it!

CONNECTIONS

The National Oceanic and Atmospheric Administration (NOAA) is the governmental body responsible for studying and reporting the weather. Media outlets analyze and summarize NOAA information for general audiences. Encourage interested youth to learn about and use NOAA weather information (www.noaa.gov).

Supplemental Information and Resources

NOAA weather information, www.noaa.gov

MINI-TORNADO

Big Picture: This activity simulates a tornado through rather simplistic means. The water in the bottle creates a vortex due to the rotation that spinning the bottle provides. As the water rotates, centripetal force pushes it toward the center. Combined with the force of gravity and the pressure gradient force pointing downward, the water is drawn into a funnel that channels it into the bottle below.

This is a little different than what actually causes a tornado, but the physics are similar. In real life, a tornado usually (but not always) derives from a parent thunderstorm called a supercell. The rotating portion of the thunderstorm, called the mesocyclone, is formed from a combination of changing wind speeds/directions with height and an updraft. As the mesocyclone rotates more quickly, sometimes a rapid downdraft (sinking of the air) allows this rotation to extend to the ground in a more intense, smaller radius. The exact mechanism for this last step is still theoretical, as more research needs to be done to confirm the theory behind it.

While the tornado in this demonstration rotates in the direction the bottle is initially spun, the vast majority of tornadoes—over 99%—spin counterclockwise (cyclonic) in the Northern Hemisphere. This is due to the Coriolis effect caused by the earth's rotation.

FACILITATING THE ACTIVITY

Youth may have trouble sealing the bottles with the tape so the washer is place correctly and no water leaks out. It's also important that the top bottle is actually spun, not just moved side to side, or the funnel will not form correctly.

SHARE WHAT HAPPENED

Question: What caused the funnel to form inside the bottle?

ANSWER:
Youth should identify that the initial spin created the funnel. They will not be able to explain the physics behind the continued motion of the vortex.

Question: How did the vortex change as time passed?

ANSWER:
Interpretive, based on observations

Question: Did the "tornado" continue until the water flowed out, or did it stop before the water ran out?

ANSWER:
It continued.

APPLY

Question: Which image is most like the vortex observed in the bottle?

ANSWER:
Youth will likely note that the first image is most similar to their funnel.

Question: What is different about the other images?

ANSWER:
The second image is wider, and the third is not a straight line from the clouds to the ground.

GENERALIZE TO YOUR LIFE

Question: If a tornado warning is issued for your area, where is the safest place to go in your school? Your home? While on the road?

ANSWER:

Home • The safest place to be when a warning is issued is inside your home (unless it is a mobile home, in which case you should leave immediately!), in the basement or lowest floor in a central room away from windows.

School • The safest place is the lowest floor, in a bathroom or central hallway away from windows.

Road • If a tornado is nearby while you're in a vehicle, do not try to dodge the tornado by driving. Get out of your vehicle and go to the nearest building or, if none is available, into a low-lying ditch.

CONNECTIONS

Unlike hurricanes, there are quite a few movies involving tornadoes. Most are highly inaccurate, unfortunately. The one that comes closest to reality is *Twister*. While the damage done by the tornadoes is fairly accurate, some of the ground action shots are unrealistic. The thunderstorms these tornadoes come from are inaccurately portrayed and do not have the structure of a typical supercell thunderstorm that would spawn tornadoes of the intensity shown in the movie.

NOAA, The U.S. National Oceanic and Atmospheric Administration, has excellent information about tornado safety:

Tornado Safety webpage, www.weather.gov/ctp/TornadoSafety

Tornado FAQs, www.spc.noaa.gov/faq/tornado/safety.html

Weather Safety webpage, www.weather.gov/safety

LIFE SKILLS

- Problem solving, critical thinking, acquiring knowledge
- Weather Skills: Gain understanding of a tornado's movement.
- Indiana Standards: 8.2.5
- NGSS: ESS2.D
- Success Indicators: Youth can see the vortex. They should also understand the difference between this vortex and a real tornado. They should know what to do if a tornado warning is issued.

SOURCES

This activity was adapted from a NOAA (National Oceanic and Atmospheric Administration/Department of Commerce) activity, https://www.noaa.gov/office-education/outreach-communication/hands-on-science-activities/tornado-in-bottle.

OUT OF THE DUST

Big Picture: This book relates a Dust Bowl story involving a young girl whose life is influenced by the plains drought of the 1930s. The drought is just a part of this story, which includes poor farming practices, the Great Depression, and a family tragedy. As this story illustrates, weather can have dramatic effects on lives, but other factors usually are involved.

FACILITATING THE ACTIVITY

The book *Out of the Dust* by Karen Hesse can be found in school or public libraries. It can also be found online at reasonable prices.

SHARE WHAT HAPPENED

Question: How did the drought and dust storms affect Billie Jo's day-to-day life?

ANSWER:

Youth should discuss the constant covering of dust, keeping plates upside down, covering silverware with napkins, stuffing doors and windows with rags, cleaning dust off her schoolwork, and having to water the apple trees. The failing crops and dying livestock also cause friends to move away and her family's poor financial situation.

Question: In any life experience, factors like the weather are only partial causes for changes. What other circumstances contributed to Billie Jo's life?

ANSWER:
The most significant event was the accidental fire, which eventually killed her mother and newborn brother as well as damaged Billie Jo's hands.

Questions: What kind of student was Billie Jo, and what were her talents? In what ways were her talents like yours?

ANSWER:
Billie Jo was a very good student and played the piano well.

Question: How were Billie Jo's teachers important?

ANSWER:
Her classroom teacher, Miss Freeland, encouraged Billie Jo to do well on the state test. Her music teacher, Arley Wanderdale, encouraged Billie Jo to play piano even after her hands were damaged in the fire.

Question: What did Billie Jo learn when she tried to run away?

ANSWER:
Billie Jo met a man who left his family to go to California, and she realized how lonely she was.

Question: What did you think was the most interesting part of the book? Why?

ANSWER:
Interpretive—no right or wrong answers

APPLY:

Question: How did the Dust Bowl change agricultural practices? Are any of these practices used today?

ANSWER:
Billie Jo's father decided to try different crops like sorghum and cotton instead of wheat.

The USDA (United States Department of Agriculture) and land-grant universities have done extensive research to improve farming practices that reduce soil erosion by both water and wind. Reduced tillage and no-till practices, improved planting practices (grassed waterways, contour planting, strip cropping), improved seed quality, removing highly erodible land from farming through Conservation Reserve Programs, and improved farming technology, all have helped reduce soil erosion. Although many parts of the Midwest experienced severe drought in 2012, the expected average national yields in 2012 would have been crop production records in 1991 due to improved farming techniques.

GENERALIZE TO YOUR LIFE

Question: Can you think of weather events—either positive or negative—that have changed things in your life or the lives of your family members?

ANSWER:
Encourage youth to think about weather events that might have changed their plans or activities, especially severe storms or long droughts.

LIFE SKILLS

- Concern for others, empathy
- Weather Skills: Weather can have significant effects on life and property.
- Success Indicator: Youth can identify significant weather events.

SUPPLEMENTAL INFORMATION AND RESOURCES

The Dust Bowl, DVD, by Ken Burns, https://www.pbs.org/kenburns/the-dust-bowl (search: The Dust Bowl)

The Grapes of Wrath, by John Steinbeck, is a story of migrants moving out of the Oklahoma area during the Dust Bowl era to find farm work in California.

SEASONS

Big Picture: Many people think that seasonal changes have to do with the changing distance between the sun and the earth. Because the earth's orbit is an ellipse and not a perfect circle, the orbit brings the earth closer to the sun; however, this occurs in January in the United States, in the middle of our winter. The earth is farthest from the sun in July, during our summer. That may not make sense, but if you lived in the Southern Hemisphere the earth would be closest during summer. The change in orbital difference is very slight, about 3% between the farthest and closest points. This change in distance has very little effect on the sun-earth relationship. Another misconception is that tilting the North Pole toward the sun brings the Northern Hemisphere closer to the sun. That logic comes from exaggerating the sun-earth model to understand the relationship. Using the scale from the activity—the sun is a basketball and the earth, a BB 74 feet away—helps to dispel the misconception of tilt bringing the Northern Hemisphere closer.

FACILITATING THE ACTIVITY

The youth might need help obtaining a globe. Ask a teacher at school or a library for permission to borrow one. The globe must be mounted on a tilted stand. Having a basketball and a BB 74 feet apart helps youth understand the magnitude of earth's orbit and the exaggeration of the models.

You might expect length of daylight at the equinox to be exactly 12 hours. We mark sunrise at the instant the sun's edge appears and sunset at the instant the sun completely disappears. So day length is 12 hours plus the time it takes for half the sun to appear on the horizon. The diagram of the sun and earth (in the Generalize to Your Life section) is typical of diagrams used to explain the seasons. But the scale of such a diagram is misleading. In this diagram the sun appears smaller than the earth. It would be difficult to have a diagram to scale, so we often exaggerate parts of it to explain how things work.

SHARE WHAT HAPPENED

Question: Do people in the Midwest receive more sunlight in the summer or winter?

ANSWER:
Indiana receives more sunlight in the summer position.

Question: During the summer is it ever dark at the North Pole?

ANSWER:
No. In the summer the North Pole is always in the light.

Question: How might these differences affect our seasons?

ANSWER:
The Northern Hemisphere receives more sunlight in the summer, which makes it warmer.-OR- The amount of sunlight changes throughout the year, and those differences cause the seasons.

Question: Contrast the difference between the summer and the winter sunlight at the North Pole.

ANSWER:
In the summer the North Pole is always in the light, and in the winter, it is always dark.

Question: Your globe might have the Tropic of Cancer marked at 23.5° north of the equator. If it does, notice that on the first day of summer the sun stays directly over that line. What line does the sun stay over on the first day of winter?

ANSWER:
On the first day of winter the sun is directly over the Tropic of Capricorn, 23.5° south of the equator.

APPLY

FIRST DAY OF SUMMER	Sunrise	Sunset	Daylight length
Indianapolis, IN	6:17 am	9:21 pm	15 hr 4 min
Miami, FL	6:30 am	8:15 pm	13 hr 45 min
Anchorage, AK	4:20 am	11:42 pm	20 hr 22 min

FIRST DAY OF FALL	Sunrise	Sunset	Daylight length
Indianapolis, IN	7:33 am	7:40 pm	12 hr 7 min
Miami, FL	7:10 am	7:15 pm	12 hr 5 min
Anchorage, AK	7:47 am	7:55 pm	12 hr 8 min

FIRST DAY OF WINTER	Sunrise	Sunset	Daylight length
Indianapolis, IN	8:06 am	5:23 pm	9 hr 17 min
Miami, FL	7:03 am	5:35 pm	10 hr 22 min
Anchorage, AK	10:14 am	3:42 pm	5 hr 28 min

Note: Summer times are daylight saving time, and winter times are standard times.

Question: Using your globe and lamp, explain why daylight lengths are different at these three locations at the summer solstice and the winter solstice.

ANSWER:
In the summer sunlight lasts longer the farther north on the globe, while in the winter sunlight lessens the farther north on the globe.

Question: Using your globe and lamp, explain why daylight lengths are similar at these three locations at the autumnal equinox.

ANSWER:
Sunlight falls equally on all parts of the globe during the autumnal equinox.

Question: What would be the daylight length on the first day of spring (vernal equinox)?

ANSWER:
Times on the first day of spring are similar to on the first day of fall.

Question: We have seasonal changes in the Midwest. Do you think seasonal changes are more or less obvious in Alaska? Explain your answer.

ANSWER:
Seasonal changes are more obvious in Alaska because the temperatures vary more than in the Midwest.

Question: Why do you think Miami, Florida, has little seasonal change?

ANSWER:
Miami has less variation in the amount of sunshine from one season to another.

GENERALIZE TO YOUR LIFE

Question: Where could you use the actual earth-sun scale to help people visualize how far we are from the sun?

ANSWER:
Youth could use a BB pellet or similarly sized item and a basketball placed 74 feet apart to show how far the earth is from the sun.

Question: How might you use scale in planning the layout of your bedroom when adding a new dresser or larger bed?

ANSWER:
Draw the outline (plan view) of a room with all the furniture in it, and use construction paper to represent the bed, dresser, chairs, and other objects because these are easier to move around than actual furniture. This can also be done in PowerPoint. Note: Measure the needed space for walkways, as this can be deceiving.

CONNECTIONS

Seasonal weather changes can affect people's ability to travel. A research station at the South Pole, the Amundsen-Scott South Pole Station, is located at the geographic South Pole on a plateau 9,301 feet above sea level. A station has been located there since 1957. More than 200 people occupy the station from mid-October through mid-February, during the Southern Hemisphere summer. In early March the sun begins to sink, and it sets completely in late March. Only 25–50 people remain for the long, cold winter night from March through October. During that time the weather is too severe to arrive or leave. A few individuals who had medical problems had to wait to get help.

LIFE SKILLS

- Acquiring knowledge
- Weather Skills: Seasons bring different weather conditions at regular cycles throughout the year caused by the earth's tilt relative to its orbit.
- Indiana Standards: 6.2.5
- NGSS: ESS1.B

- Success Indicator: Youth recognize that seasons are caused by the tilt of the earth and the resulting change in the amount of sunlight earth receives during each season; and that the severity of seasonal changes increases at higher latitudes.

SUPPLEMENTAL INFORMATION AND RESOURCES

Sunrise/sunset times: www.timeanddate.com/worldclock/sunrise.html

USING WEATHER INSTRUMENTS TO COLLECT WEATHER DATA

Big Picture: Weather greatly affects our daily life and activities. Farmers and gardeners depend on favorable weather to grow and harvest crops. Severe weather disrupts many aspects of our daily lives. Crop damage from weather conditions increases the cost of food. Transportation is impacted when planes are grounded and icy highways cause accidents. Snow can make buses late or even give students a day off of school! Ice can damage telephone and electric lines. Recent major weather events include Hurricane Katrina in New Orleans and the surrounding area in 2005 and an extended drought in the Midwest in 2012.

Weather Service forecasts under these circumstances may save thousands of lives and millions of dollars in property damage. Weather forecasts warn Florida fruit growers of early frosts in time to set out smudge pots. Ranchers who are warned of storms can provide shelter and food for cattle. Flood and tornado warnings save lives and property.

SHARE AND APPLY QUESTIONS

ANSWER:
Interpretive-no right or wrong answers

GENERALIZE TO YOUR LIFE

Question: What can you learn by averaging the weather data from 30 years of records?

ANSWER:
A 30-year average is the usual method used to determine climate data. Averaging data from 30 years gives a good idea of the weather conditions that can be expected for an area. Weather can and does change from day to day—a common comment in Indiana is, "If you don't like the weather now, wait a minute, it will change"—but over the long term, it is fairly consistent in an area. Long-term data shows that many places are warming at an accelerated rate (over historic increases), which is generally attributed to human influences. The glaciers, for example, are melting at a much higher rate than has ever occurred.

ACTIVITY FACILITATION AND ANSWERS: LEVEL 3

AIR MASSES AND FRONTS

Big Picture: It is important to understand that fronts result from movement of air masses. The air masses move because of differences in air pressure. Specifically, low-pressure areas pull air masses in and around the rotational motion created by the low. This activity is closely linked to the High and Low Pressure Systems activity but can be done independently of it. The severity of any front depends on the differences between the involved air masses. The greater the difference in temperature and moisture, the more severe the resulting weather. The strength of the low pressure is also a factor. The greater the difference in pressure, the stronger the winds and the more severe the weather. All of these factors influence the type of weather that results.

FACILITATING THE ACTIVITY

Match the air masses to their characteristic:

Air Mass
A. Maritime Tropical (mT)
B. Continental Tropical (cT)
C. Continental Polar (cP)
D. Maritime Polar

Letter	Characteristic
B	Dry and warm
D	Moist and cold
A	Moist and warm
C	Dry and cold
B	Forms over warm land
A	Forms over warm water
D	Forms over cold water
C	Forms over cold land

The cold front

Question: What causes the movement of air masses?

ANSWER:
Pressure differences cause air mass movement.

Question: What causes the warm air ahead of a cold front to rise?

ANSWER:
The colder, denser air forces the warm air to rise. OR A difference in air density.

Question: What kind of clouds usually form in a cold front?

ANSWER:
Towering cumulus and cumulonimbus clouds.

Question: What wind direction change usually occurs as a cold front moves through an area?

ANSWER:
A change in wind direction from southwest to northwest usually occurs.

Question: Where does vertical wind shear form?

ANSWER:
Vertical wind shear occurs at the frontal boundary.

Question: What kind of weather is usually expected in a cold front?

ANSWER:
Stormy weather.

The warm front

Question: What cloud types are typically associated with a warm front?

ANSWER:
Stratus and nimbostratus near the frontal boundary, with cirrostratus and cirrus clouds in advance of the front.

Question: Which air mass provides the moisture that produces clouds and precipitation?

ANSWER:
The warm air mass provides the moisture.

Question: What severe events could warm fronts produce?

ANSWER:
Ice storms, blizzards, and heavy rains that cause floods.

Question: What kind of weather is usually expected in a warm front?

ANSWER:
Overcast skies and precipitation lasting for long periods.

Question: How does an occluded front form?

ANSWER:
An occluded front forms when a cold front overtakes a warm front.

Question: Within the low, where do occluded fronts form?

ANSWER:
Occluded fronts form near the center of the low.

Question: How do stationary fronts form?

ANSWER:
Stationary fronts form when a cold or warm front stops moving across the surface.

Question: Within the low, where do stationary fronts form?

ANSWER:
Stationary fronts form away from the center of the low.

SHARE WHAT HAPPENED

Question: What causes stormy weather in the Midwest?

ANSWER:
Cold fronts cause stormy weather.

APPLY

Question: When you see stratus and nimbostratus clouds, what weather would you predict?

ANSWER:
Expect a warm front, which forms clouds that cover a large area and create rain or snow events that can last for days.

GENERALIZE TO YOUR LIFE

Question: How can you use the information you learned to help you understand TV weather reports?

ANSWER:
When a weather report mentions that a cold front is approaching, youth should expect stormy weather and consider changing their plans or taking appropriate clothing.

CONNECTIONS

Vertical wind shear is dangerous for airplanes, particularly at takeoff and landing. Airplanes cannot adjust for this rapid change in airflow. During takeoff, the pilot is trying to achieve as much lift as possible. During landing, the pilot is slowing the plane and trying to level the plane to the surface. In both cases the plane is close to the ground, and the pilot has no time to adjust to a sudden change in the vertical movement of air, which could cause a disaster. Pilots do not attempt takeoffs and landings when a cold front is nearby. Controllers use Doppler radar to detect wind shears and warn pilots of them.

LIFE SKILLS

- Acquiring knowledge, critical thinking
- Weather Skills: Understand the development of air masses and the resulting formation of fronts and the weather those fronts produce.
- Indiana Standards: ES.4.4
- NGSS: ESS2.D
- Success Indicator: Youth can explain how air masses form, and how these air masses form fronts with the resulting types of clouds/weather.

ISAAC'S STORM

Big Picture: The author of *Isaac's Storm*, Erik Larson, does more than tell the story of the devastating Galveston hurricane of September 8, 1900. He also describes the science behind the development of a severe hurricane. He discusses the history of understanding how weather works. He cites the experiences of early explorers encountering tropical storms and hurricanes. He relates the political struggles of the early Weather Bureau. Most importantly he tells the story of Isaac Cline's move from a farm in Tennessee to head of the Texas Weather Bureau office in Galveston.

FACILITATING THE ACTIVITY

This book can be found in school or public libraries and online at reasonable prices.

SHARE WHAT HAPPENED

Question: Why was Galveston an important city in Texas?

ANSWER:
Galveston was the most important port in Texas and was vying with Houston to become the state's greatest commerce center.

Question: What was the general opinion of Galveston's safety from a hurricane in 1900?

ANSWER:
Galveston was thought to be safe from tropical storms. Isaac Cline had published a paper showing how tropical storms and hurricanes would never reach the Galveston coastline.

Question: How did Isaac Cline come to be the head of the Galveston weather station?

ANSWER:
Isaac was recruited into the Weather Bureau when it was still part of the Army Signal Corps. He was assigned to the Abilene, Texas, station, where he met his wife. From there he was named head of the Texas Weather Bureau headquartered in Galveston.

Question: Why did the Weather Bureau ignore reports from Havana, Cuba?

ANSWER:
The Weather Bureau did not trust local reports from Cuba and thought they were often exaggerated. The Weather Bureau did not want to unnecessarily panic the public.

Question: What is the significance of the ring on Isaac's finger in his photograph?

ANSWER:
The ring that Isaac wears in the photograph belonged to his wife, who died during the hurricane.

Question: What part of this book was most interesting to you?

ANSWER:
Interpretive—no right or wrong answer.

APPLY

Question: How would a hurricane like this be handled differently today?

ANSWER:
Today we have much better technology, and the National Weather Service broadcasts storm watches and warnings well in advance of a storm's arrival. People have time to prepare or even to be evacuated from the area if necessary. The loss of life would be significantly lower.

GENERALIZE TO YOUR LIFE

Question: Can you think of weather events either positive or negative that have changed things in your life?

ANSWER:
Encourage youth to think about weather events that may have changed their plans or activities.

Question: Can you think of weather events either positive or negative that have changed things in the lives of any family members?

ANSWER:
Encourage youth to think about weather events that may have changed their plans or activities.

LIFE SKILLS

- Concern for others; personal safety
- Weather Skills: Weather can have significant impact on life and property.
- Indiana Standards: ES.4.4
- NGSS: ESS2D, ESS3B
- Success Indicator: Youth can identify significant weather events.

MONITORING WEATHER

Big Picture: Youth are encouraged to become a CoCoRaHS Network volunteer and collect and record (online) precipitation data for one year. This is a big commitment, possibly beyond the interest of most youth due to the commitment beyond the typical "4-H year." Family members or friends can share this activity to reduce the burden of regular reporting by one person.

FACILITATING THE ACTIVITY

Attending training helps youth understand CoCoRaHS's goals and the data collection techniques and methods it requires. It also connects them with an expert and other volunteers.

SHARE WHAT HAPPENED

Questions: from NOAA

ANSWERS:
Youth's answers will vary depending on their experiences and insight. Interpretive—no right or wrong answers.

LIFE SKILLS

- Keeping records; volunteering; community service
- Weather Skills: Collecting accurate data; sharing data
- NGSS: none
- Success Indicator: Youth collect and record precipitation data at the CoCoRaHS website.

PRESSURE SYSTEMS

Big Picture: High and low pressure systems begin to form due to pressure differences in the atmosphere. The initial movement of air is the atmosphere's response to equalize pressure differences. The Coriolis effect, caused by the earth's rotation, moves air currents to the right in the Northern Hemisphere and to the left in the Southern Hemisphere. Friction reduces the Coriolis effect, allowing the wind to eventually establish equilibrium. Sinking air in a high pressure system produces clear weather. Rising air in a low pressure system, and the formation of fronts around the low, cause cloudy, rainy weather.

FACILITATING THE ACTIVITY

Any round container with a flat bottom works for the activity. The larger the pan, the more dramatic the results.

Pie pan activity

Question: What did you observe when the red sugar water was dropped into the center of the pan?

ANSWER:
The sugar water sank to the bottom and began to spread out.

Questions: What did you observe when you rotated the pan?

ANSWER:
The sugar water moved to the outside of the pan.

Questions: What did you observe after the pan was still again?

ANSWER:
The sugar water moved back toward the center of the pan.

Questions: How does this activity help explain pressure systems?

ANSWER:
The sugar water is denser than the warm water in the pie pan. Dropping the sugar water into the middle of the pan creates a high-pressure area, so the sugar water should sink and begin to disperse. Adding more sugar water increases the size of the high-pressure area. A centrifugal force (a force that tends to impel something outward from a center of rotation) is created when the youth rotates the pan. The centrifugal force pushes the denser sugar water to the outside of the pan. The center of the pan then fills up with the less dense clear water, creating a low-pressure area in the center of the pan. When the pan is still again, youth should observe the denser sugar water moving back toward the center.

Development of high and low pressure systems

Question: Which is denser, warm or cold air?

ANSWER:
Cold air

Question: Which is denser, moist air or dry air?

ANSWER:
Dry air

Question: Which air forms a high pressure system, denser or less dense air?

ANSWER:
Denser air

Question: Which air forms a low pressure system, denser or less dense air?

ANSWER:
Less dense air

Question: The Coriolis effect in the Northern Hemisphere causes wind to be deflected to the right or left?

ANSWER:
To the right

Question: What is the resulting rotation of a high, clockwise or counterclockwise?

ANSWER:
Clockwise

Question: What is the resulting rotation of a low, clockwise or counterclockwise?

ANSWER:
Counterclockwise

Question: How is friction involved in the formation of a high or low?

ANSWER:
Friction reduces the Coriolis effect, allowing the air to move from the high to the low.

Question: Why do few clouds form in a high?

ANSWER:
Few clouds form in a high because the air is sinking.

Question: Why are clouds expected to form in a low?

ANSWER:
Air rises and fronts form in a low.

Question: What kind of weather is expected in a high?

ANSWER:
Clear weather is expected in a high.

Question: What kind of weather is expected in a low?

ANSWER:
Cloudy, rainy weather is expected in a low.

SHARE WHAT HAPPENED

Questions: What was the most interesting part of this activity?

ANSWER:
Interpretive—no right or wrong answer.

APPLY

Question: Without repeating the experiment above, tell another way you could teach someone your age about the structure and development of high and low pressure systems and their effect on our weather.

ANSWER:
Youth might use their hands or a drawing to show how one front can replace another.

GENERALIZE TO YOUR LIFE

Question: How can you use the information you learned to help you understand TV weather reports?

ANSWER:
Understanding pressure systems helps youth read weather maps and predict coming weather.

CONNECTIONS

High and low pressure systems in the Southern Hemisphere form differently than in the U.S. because the Coriolis effect deflects wind to the left in the Southern Hemisphere. As a result, lows rotate clockwise, and highs rotate counterclockwise. Although rotation is different, the type of weather is the same. Highs produce clear weather, and lows produce cloudy weather.

LIFE SKILLS

- Problem solving
- Weather Skills: Understand the formation of high and low pressure systems.
- Indiana Standards: ES.4.2
- NGSS: ESS2.D
- Success Indicator: Youth can explain how highs and lows develop and can describe the weather associated with these systems.

SUPPLEMENTAL INFORMATION AND RESOURCES

High-low image: https://www.wpc.ncep.noaa.gov/archives/web_pages/discussions/archive_nathilo.php

Weather map image: https://www.nesdis.noaa.gov/about/k-12-education/weather-forecasting/how-read-weather-map

WEATHER STATION MODELS

Big Picture: Today's technology makes it relatively easy for scientists to track the movement of storms. Watching how a storm develops and moves provides insight into how meteorologists watch the weather. Youth are asked to track the movement of a current storm, rather than one that has already occurred, so they can try looking at the data as it develops and making their own predictions.

Over time, the thousands of weather stations around the world have undergone changes that often result in sudden or unrealistic discrepancies in observed temperatures requiring a correction. For the U.S.-based stations, we have access to detailed station history that helps us identify and correct discrepancies. Some of these differences have simple corrections.

The most important difference globally was the modification in measured sea surface temperatures. In the past, ship measurements were taken by throwing a bucket over the side, bringing some ocean water on deck and putting a thermometer in it. Today, temperatures are recorded by reading thermometers in the engine coolant water intake—this is considered a more accurate measure of ocean temperature. The bucket readings used early in the record were cooler than engine intake observations, so the early data have been adjusted warmer to account for that difference. This makes global temperatures indicate less warming than the raw data does.

The most important difference in the U.S. temperature record occurred with the systematic change in observing times from the afternoon, when it is warm, to morning, when it is cooler. This shift has resulted in a well-documented and increasing cool discrepancy over the last several decades and is addressed by applying a correction to the data."

Q&A from NOAA, www.noaa.gov

FACILITATING THE ACTIVITY

Question: Using data and weather maps, encourage youth to make a comprehensive study of a significant storm event as it happens. They should also include newspaper and Internet articles related to the storm to better display its impact.

ANSWER:
Open-ended—no right or wrong answers.

SHARE WHAT HAPPENED

Question: How many weather symbols can you draw without referring to any reference?

ANSWER:
Interpretive—no right or wrong answers.

APPLY

Question: How might you use your station model sequence to predict future weather? What additional information do you need?

ANSWER:
Some people keep weather data for many years, essentially creating their own microclimate record for the exact spot where they live. Data from 20–30 years is necessary to predict future weather accurately.

GENERALIZE TO YOUR LIFE

Question: Why are symbols and acronyms useful?

ANSWER:
Acronyms help us refer to a favorite university or a governmental agency quickly.

Question: What other symbols can you list?

ANSWER:
Symbols are used in many areas, including on road signs and in facial expressions, sign language, and company logos.

Question: What other acronyms can you list?

ANSWER:
Open-ended—no right or wrong answers.

LIFE SKILLS

- Learning to learn; keeping records
- Success Indicator: Youth can use data resources to better understand the movement of storms.

SUPPLEMENTAL INFORMATION AND RESOURCES

NOAA Surface station model map or analyzed station model map: www.spc.noaa.gov/obswx/maps/

NOAA current severe weather map: www.spc.noaa.gov/climo/reports/

Current wind map: http://hint.fm/wind/

Weather map station model images: https://www.wpc.ncep.noaa.gov/html/stationplot.shtml

While NOAA sources are preferred, you can share the following websites with youth if you wish:

The National Weather Service, www.weather.gov

The Weather Channel current interactive weather maps: After entering your zip code in the search box on the left, you can scroll down to find the hourly weather forecast and radar and satellite images.

WEATHER IN THE TROPOSPHERE

Big Picture: The troposphere is the layer of air closest to the surface of the earth. It is on average seven miles thick but contains almost 90% of the earth's atmosphere. Weather takes place in the troposphere. The surface of the earth, not the sun, warms the troposphere. The resulting global wind system is also in the troposphere. These patterns are affected by the seasons as well as differences in the earth's surfaces. Jet streams, which are rivers of fast-moving air, exist at the top of the troposphere in between the different sections or cells of the global wind pattern. The polar jet stream has a strong influence on the type of weather we experience in Indiana. It fluctuates north or south depending on several factors at the surface. Abnormal warming of the oceans can greatly affect these weather patterns.

Atmospheric particulate matter, also known as particulates or particulate matter (PM), are tiny pieces of solid or liquid matter associated with the earth's atmosphere. They are suspended in the atmosphere as atmospheric aerosol, a term that refers to the particulate/air mixture. Sources of particulate matter can be manmade or natural. They can adversely affect human health and also impact climate and precipitation. For more information see www.epa.gov/pm-pollution/particulate-matter-pm-basics.

FACILITATING THE ACTIVITY

Having some background in global wind patterns, high and low pressure areas, and the nature of weather fronts makes this activity easier to understand. Encourage youth to review those activities before proceeding.

The structure of the atmosphere

Question: What is the trend of temperature change as altitude increases in each layer?

ANSWERS:

Thermosphere: increase

Mesosphere: decrease

Stratosphere: increase

Troposphere: decrease

Question: Why do you think temperatures increase in the thermosphere?

ANSWER: High-energy radiation like x-rays and UV light is absorbed in the thermosphere.

Question: Why do you think temperatures increase in the stratosphere?

ANSWER:

The ozone in the stratosphere absorbs UV light.

The troposphere

Question: Why do mountain climbers sometimes need oxygen tanks?

ANSWER:

A breath of air on a mountain has only about 30% of the oxygen in a breath of air at sea level.

Question: Why is the temperature greatest at the surface of the earth?

ANSWER:

The air in the troposphere absorbs very little of the sun's radiation. The sun must warm the surface of the earth, which in turn warms the air.

Question: If it is 65°F at sea level and you are on the top of Mount Everest, what would you expect the air temperature to be, based on altitude?

ANSWER:

Calculations: 5.5 miles * 5,280 ft/mile * 3.3 degrees/1,000 ft = 29.04 degrees different

65 – 29 = 36°F

The jet stream

Question: How were the jet streams discovered?

ANSWER:

In 1944, during WWII, U.S. pilots in a bombing run over Tokyo discovered an area of high winds that turned out to be a jet stream.

Question: Where is the polar front located?

ANSWER:

The polar front is located at the top of the troposphere, about 60° latitude between the Westerlies (Ferrel Cell) and the Polar Easterlies.

Question: Why is the wind speed highest in the center of a jet stream?

ANSWER:

Frictional forces are low in the center. Friction away from the center slows the outside air down.

The polar jet stream and climate change

Question: Why does the polar jet stream shift northward during our summer?

ANSWER:

The sun is higher and warms the Northern Hemisphere more, which shifts the global wind pattern northward.

Question: Why is cold-front weather more severe in the spring and summer in Indiana?

ANSWER:
The polar jet stream tracks low-pressure areas north of Indiana.

Question: Why are warm fronts more severe in the fall and winter in the Midwest?

ANSWER:
The polar jet stream tracks low-pressure areas south of Indiana.

Question: How can El Niño and La Niña affect the polar jet stream?

ANSWER:
El Niño forces the polar jet stream northward, and La Niña forces it southward.

Question: What will the polar jet stream do when the North Atlantic Ocean remains warm?

ANSWER:
A warm North Atlantic blocks the polar jet stream and forces it to plunge into the eastern part of the U.S.

Question: What kind of weather can be expected in the eastern half of the U.S. when there is a polar jet stream blocked by warm North Atlantic temperatures?

ANSWER:
This causes cold winter weather in the southeastern states and heavy snowfall along the northeastern coast.

SHARE WHAT HAPPENED

Question: How is a jet stream like a river?

ANSWER:
A jet stream contains air that flows in a tubular motion and is fastest in the center. Jet streams are large enough to contain multiple airplanes.

APPLY

Questions: How could you use what you learned in this activity to improve your study of weather? How might you use it in your everyday life?

ANSWER:
This activity helps youth visualize large weather patterns and the interactions among multiple influences (e.g., gravity, jet streams, low pressure systems). They can use this information to gain a deeper understanding of weather reports and how weather determines climate.

GENERALIZE TO YOUR LIFE

Question: How is climate change affecting the weather in the troposphere?

ANSWER:
Burning fossil fuels causes increased concentrations of greenhouse gases, which trap outgoing radiation. This is causing average earth temperatures to rise more rapidly than has ever occurred before. More frequent extreme patterns are bringing more extreme weather.

LIFE SKILLS

- Learning to learn
- Weather Skills: The ability to recognize components of the troposphere that create weather.
- Indiana Standards: ES.4.1, ES.4.3
- NGSS: ESS2, HS-ESS2-4
- Success Indicator: Youth understand the dynamics of weather that occurs within the troposphere.

SUPPLEMENTAL INFORMATION AND RESOURCES

NOAA Weather School, Layers of the Atmosphere, www.noaa.gov/jetstream

Layers of the Earth's Atmosphere, Windows to the Universe (National Earth Science Teachers Association), www.windows2universe.org/Earth/Atmosphere/layers.html

Snow cover extent (northern hemisphere), www.ncei.noaa.gov/products/climate-data-records/snow-cover-extent

Britannica Kids, wave cyclone, http://kids.britannica.com/comptons/art-54138/In-the-early-stages-of-a-wave-cyclone-an-eastward

WINDCHILL AND HEAT INDEX

Big Picture: "Feels-like" temperatures are not true temperatures. Windchill temperatures cool the body as their equivalent temperature would and decrease the time it takes for frostbite or hypothermia to set in. The heat index, based on the amount of moisture in the air, is similar. With more moisture in the air, perspiration cannot evaporate as quickly from skin, shortening the time it takes for severe heat conditions to occur.

FACILITATING THE ACTIVITY

Surface and wind

Question: Do the surface temperatures feel the same on the towel and the cookie sheet?

ANSWER:
No.

Question: Which surface feels cooler?

ANSWER:
The cookie sheet feels cooler.

Question: How does the air blowing on your right hand feel compared to your left hand, which is not in front of the fan: cooler (expected), the same, or warmer?

ANSWER:
Cooler

Question: Did the fan change the temperature of a thermometer at room temperature?

ANSWER:
No.

Note: Results will vary, but there will be a significant difference in the cooling time. Youth are asked to repeat this activity five times because conditions can vary slightly as can human actions (timing and viewing the thermometer) .

Heating and Cooling Results Chart: Time for a 10°F Drop in Temperature. Examples in red:

Repetition (Rep)	Without a fan (seconds)	With a fan (seconds)
1	45	15
2	43	14
3	48	16
4	44	15
5	45	14
Average	45	14.8

Question: How much faster did the thermometer cool when the fan was blowing on it? Why?

ANSWER:
In the example, the fan cooled the thermometer about three times faster (45/14.8). This is because the fan moves the warm air away from the thermometer (convection cooling).

What it feels like

Question: Why do you think that is? (Why does the cookie sheet feel cooler than the towel, even though both are the same temperature?)

ANSWER:
Metal conducts heat much more quickly than the towel does. Although both items are at room temperature, the youth's body temperature is lower, so it feels like heat is being lost to the cookie sheet.

The reason

Question: Scientifically speaking, why should you use a towel or hot pad to grab a hot cookie sheet?

ANSWER:
Heat transfers from the cookie sheet faster than it transfers through a towel or hot pad, which protects your hand.

Windchill

Question: What is the windchill at 40°F with a wind speed of 5 mph?

ANSWER:
36°F

Question: What is the windchill at 40°F with a wind speed of 25 mph?

ANSWER:
29°F

Question: Will water freeze when the air temperature is 35°F with a windchill that is lower than 32°F?

ANSWER:
No.

Heat index

Question: What is the heat index at 82°F and 50% relative humidity?

ANSWER:
83°F

Question: What is the heat index at 82°F and 85% relative humidity?

ANSWER:
88°F

Question: In Arizona the temperature is 88°F with a relative humidity of 45%. What is the heat index? What is the danger level for strenuous activity?

ANSWER:
89°F. Caution.

Question: In Indiana on the same day the temperature is 88°F with a relative humidity of 75%. What is the heat index? What is the danger level for strenuous activity?

ANSWER:
103°F. Danger.

Question: People say that Arizona has a dry heat. What do you think this means?

ANSWER:
With less humidity, the temperature does not feel as hot because water evaporates better.

Question: Why is it important to drink plenty of water on days with a high heat index?

ANSWER:
Drink plenty of water so you can sweat, which helps cool you. Water is best; many drinks, including those that contain caffeine or alcohol, act as diuretics and promote urine production.

Question: What can you do to protect yourself in severe windchill situations (examples):

ANSWERS:
Wear layered clothes.
Wear a hat that covers the ears.
Wear gloves and a scarf to cover your face.
Stay indoors.

Question: What can you do to protect yourself in severe heat situations (examples):

ANSWERS:
Wear light clothing.
Go swimming.
Stay in shaded areas.
Stay indoors with air conditioning.

SHARE WHAT HAPPENED

Question: How close were your five readings in the Heating and Cooling Results Chart?

ANSWER:
Time measurements should be within a few seconds for each column. Encourage youth who are interested in learning more to calculate the standard deviation of their data. The standard deviation is calculated by:

- **a** Calculate the average (sum the times/5).
- **b** Calculate the difference between each reading and the mean, and square this value.
- **c** Add the values calculated in **b** (sum the 5 squared differences).
- **d** Divide the sum from **c** by 5 (the number of data points).
- **e** Take the square root of the number found in **d**.

A low standard deviation indicates that the data points tend to be very close to the average value (expected value). A low standard deviation is a goal that shows good replication of methodology.

APPLY

Question: Why might a room feel cool when one or two people are in it but warm when 100 people are in it, even though the air-conditioning setting has not changed?

ANSWER:
People give off heat. Many people in an air-conditioned room can produce more heat than the air conditioner can handle.

Question: What did you learn from this activity that can make you safer?

ANSWER:
Youth should understand the importance of covering their skin when the windchill reflects cold temperatures; and staying out of the sun, drinking plenty of water, and finding a fan or breeze when the heat index is high.

GENERALIZE TO YOUR LIFE

Question: Why is it important for parents to understand the windchill and heat index temperatures?

ANSWER:
Understanding these temperature issues could be lifesaving for mammals—humans and animals—in extreme weather conditions.

LIFE SKILLS

- Critical thinking; keeping records; learning to learn
- Weather Skills: Understand and determine feels-like temperatures.
- Indiana Standards: ES.4.4
- NGSS: ESS2.D
- Success Indicator: Youth can determine windchill and heat index temperatures.

SUPPLEMENTAL INFORMATION AND RESOURCES

From NOAA (National Oceanic and Atmospheric Administration):

Please change to NOAA Heat forecast tool, new in 2024: www.noaa.gov/news-release/noaa-expands-availability-of-new-heat-forecast-tool-ahead-of-summer

Windchill Chart: www.noaa.gov/jetstream/synoptic/wind-chill

Heat Index Calcuation: www.wpc.ncep.noaa.gov/html/heatindex.shtml

Global winds website, http://Earth.nullschool.net/ (viewpoint can be changed)

CLIMATE AND CLIMOGRAPHS

Big Picture: Climate is largely determined by a region's precipitation and temperature. Climate norms are based on 30-year periods. Current norms are based on data from 1991 through 2020 (reference: wmo.int/media/news/updated-30-year-reference-period-reflects-changing-climate). Climographs show the relationship of temperature and precipitation in a location and are a convenient way to compare different locations.

FACILITATING THE ACTIVITY

Indianapolis climograph

Question: What is the warmest month in Indianapolis, according to the climograph?

ANSWER:
July.

Question: What is the wettest month?

ANSWER:
June.

Question: Which months are best for growing crops?

ANSWER:
May–September (answers may vary)

Question: How does the Indianapolis growing season compare to the growing season where you live?

ANSWER:
Open-ended.

Question: During which months is the average temperature in Indianapolis below freezing (32°F)?

ANSWER:
November through April.

Question: Do you think stores sell more long underwear or bathing suits annually in Indianapolis?

ANSWER:
Open-ended, but youth might say long underwear, because there are more cold months than warm months.

U.S. climographs

Question: Which city receives the most precipitation?

ANSWER:
C

Question: Which city receives the least precipitation?

ANSWER:
B

Question: Which city has the highest temperatures?

ANSWER:
B

Question: Which city has the lowest temperatures?

ANSWER:
F

Question: Which city has the most distinct change in seasons?

ANSWER:
A, D, F (interpretive)

Question: Which city has the least changes in seasons?

ANSWER:
B

Question: Circle what you would expect to have most of in your closet to be prepared for the weather for each city: (answers may vary slightly)

City	More shorts or pants?	More sandals or shoes?	More raincoats or sunscreen?	More heavy coats or light jackets?
A	pants	shoes	raincoats	light jackets
B	shorts	sandals	sunscreen	light jackets
C	pants	shoes	both	light jackets
D	pants	shoes	raincoats	light jackets
E	pants	shoes	raincoats	light jackets
F	pants	shoes	raincoats	heavy jackets

Question: Match the cities:

ANSWERS:

A Boston, Massachusetts

B Phoenix, Arizona

C Miami, Florida

D Chicago, Illinois

E Portland, Oregon

F Duluth, Minnesota

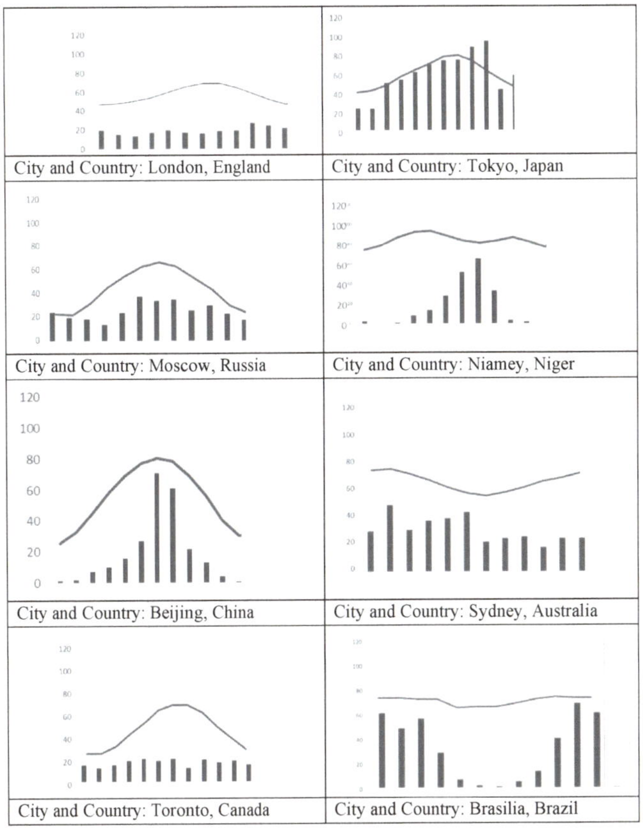

SHARE WHAT HAPPENED

Question: What does a climograph show?

ANSWER:
A climograph is a visual representation of a location's temperature and precipitation. Climographs give us an idea of the type of weather expected there.

APPLY

Question: What can you learn by comparing climographs from different cities?

ANSWER:
Climographs give a quick overview of a city's climate, both the average temperature and rainfall. This data can help youth understand how people might dress or the types of crops that can be grown.

GENERALIZE TO YOUR LIFE

Question: How might the climographs that you studied in this activity change over the next 50 years?

ANSWER:
Temperatures are expected to increase in many areas, particularly those with a lot of impervious surfaces. Severe weather such as droughts or storms are expected to increase. The effect of climate change on precipitation has not been studied in as much detail as changing weather.

LIFE SKILLS

- Critical thinking; problem solving
- Weather Skills: Recognize the importance of a climate's precipitation and temperature patterns.
- Indiana Standards: ESA.3.3
- NGSS: ESS2.D, ESS3.D
- Success Indicator: Youth can read and construct climographs, and from them, predict climate patterns for locations and understand how they influence living conditions.

DROUGHT MONITORING

Big Picture: Drought can have many consequences. The NOAA Drought Monitor website publishes weekly drought maps and data on the national and state levels. These are archived and easily accessible. During any serious drought, news organizations and government websites also issue reports.

FACILITATING THE ACTIVITY

This activity is set up as an inquiry in which youth choose a drought situation, then display the progression of data that shows its length and severity. They might need help deciding which data to display to best illustrate the extent of the drought conditions; and transforming data into graphic displays with graph paper or Excel. Encourage youth to be comprehensive in their research into the drought's consequences.

The display or report should include graphic data of a region's drought episode. It could also include maps and charts. Consequences of the drought may be presented in pictures, displayed data, and narratives of agricultural, environmental, economic, and social effects. All information needs to be cited.

APPLY

Question: How might climate change affect drought monitoring?

ANSWER:
Most climate scientists expect droughts to increase both in length and severity in coming years. Drought occurs when a region receives less than normal rainfall to the extent that there is not enough water to sustain normal conditions. Normal is based on a 30-year data average, so if an area has a sustained decrease in rainfall over many years, the 30-year average decreases, and a new normal is established.

GENERALIZE TO YOUR LIFE

Question: How could an increase in the incidence of droughts affect your landscaping plans for the first home you purchase?

ANSWER:
Our changing climate is expected to cause sustained changes in weather conditions. As a result, homeowners (and farmers) might have to reconsider the species they can grow successfully, and when they can plant and harvest.

Some invasive species and crop pests also appear to be moving northward into areas that were once too cold for them. These pests need to be managed.

LIFE SKILLS

- Learning to learn; keeping records
- Weather Skills: Recognize drought conditions and the effect drought can have on a region's environment, agriculture, and economics.
- Indiana Standards: ES.4.4
- NGSS: ESS2.D
- Success Indicator: Youth use the U.S. Drought Monitor to recognize drought episodes and research the agricultural, environmental, economic, and social effects of drought on a region.

SUPPLEMENTAL INFORMATION AND RESOURCES

Indiana's Water Shortage Plan: www.in.gov/dnr/water/water-availability-use-rights/indianas-water-shortage-plan/

United States Drought Monitor: www.droughtmonitor.unl.edu

ENERGY IN THE ATMOSPHERE

Big Picture: Understanding radiation, conduction, and convection is essential to understanding the dynamics of the atmosphere. No form of radiation affects nitrogen and oxygen molecules. Nitrogen and oxygen make up 99% of the atmosphere. Greenhouse gases are not affected by sunlight (shortwave radiation), but do absorb the infrared radiation (longwave radiation) emitted by the warm surface of the earth. This minor part of the atmosphere is responsible for keeping the atmosphere at tolerable temperature levels. Increases in these greenhouse gases raise the atmosphere's temperature.

FACILITATING THE ACTIVITY

Radiation

Question: Describe what you felt when your hand was beside the flame.

ANSWER:
Youth should feel warmth from the candle when their hand is about 1 inch from the flame.

Question: How far was your hand from the flame when you could no longer feel the heat?

ANSWER:
Youth should not be able to feel heat from the candle when their hand is 3 inches away from the flame.

Conduction

Question: What happened to the paper clip in the flame?

ANSWER:
The paper clip heats up.

Question: What happened when you touched it to the paper?

ANSWER:
The paper clip scorches the paper. It might burn a hole in the paper but does not set it alight.

Convection

Question: About how far was your hand from the flame when you first felt the heat?

ANSWER:
Youth should feel warmth from the candle at about 10 inches above the flame. If they move their hand down another inch or two they should feel much more heat than when their hand was only 1 inch to the side of the flame. Encourage them to think about why this might be.

Question: What did you feel when your hand was above the flame?

ANSWER:
The warmer, less dense air is rising. This is called convection. This experiment illustrates how more heat is transferred by convection than by radiation.

Matching

C—convection: transfer of energy by movement in a gas or liquid

B—conduction: transfer of energy that requires contact

A—radiation: transfer of energy as waves through space

Energy from the sun

Question: As radiation strikes a substance, what are the three possible results?

ANSWER:
Radiation can pass through the substance, be reflected, or be absorbed into the substance. Combinations are also possible.

Question: Which do you think will absorb more sunlight, snow or concrete?

ANSWER:
Concrete absorbs more sunlight. Snow reflects quite a bit of sunlight.

Albedo

Question: Which will reflect more sunlight, a substance with a low albedo or high albedo?

ANSWER:
A substance with a high albedo—snow, for example—reflects more sunlight.

Question: What happens to reflected sunlight?

ANSWER:
Reflected sunlight is lost back to space unless it is captured by clouds or greenhouse gases.

Energy in the atmosphere

Question: How does sunlight affect the molecules in the atmosphere?

ANSWER:
Sunlight has no effect on the molecules in the atmosphere.

Question: How does infrared radiation affect oxygen and nitrogen molecules?

ANSWER:
Infrared radiation has no effect on the molecules in the atmosphere.

Question: What percentage of the atmosphere are the greenhouse gases?

ANSWER:
Less than 0.5%.

Question: What might a rise in greenhouse gases do to the temperature of the atmosphere?

ANSWER:
A rise in greenhouse gases would increase the temperature.

SHARE WHAT HAPPENED

Question: Name the three main types of energy transfer.

ANSWER:
Radiation, convection, conduction.

APPLY

Question: Why can it be helpful to wear light-colored clothes on a hot, sunny day and dark-colored clothes on a cold, sunny day?

ANSWER:
Wearing light-colored clothes in the summer keeps someone cooler than dark colors because of the albedo effect. Similarly, wearing darker clothes in the winter helps keep someone a little warmer than wearing lighter clothes.

GENERALIZE TO YOUR LIFE

Question: How might the albedo, or reflecting power, of an area affect your comfort?

ANSWER:
Understanding radiation, convection, and conduction help youth figure out why is it warmer near a stove or a fire but not a short distance away; and why a warm breeze moves heat in the summer.

LIFE SKILLS

- Learning to learn
- Weather Skills: The energy that powers weather comes from the sun.
- Indiana Standards: ES.3.2, ES.4.2
- NGSS: ESS2.D, ESS3.D
- Success Indicator: Youth describe how radiation, conduction, and convection work to transfer energy throughout the earth system.

SUPPLEMENTAL INFORMATION AND RESOURCES

Carbon dioxide information: https://gml.noaa.gov/ccgg/trends/

Albedo information: https://www.ospo.noaa.gov/products/land/lsa/#:~:text=Surface%20Albedo%20(SURFALB)%2C%20defined,surface%20and%20the%20climate%20system

Atmospheric gases information: https://www.noaa.gov/jetstream/atmosphere

Radiation information: https://www.ncei.noaa.gov/products/climate-data-records/outgoing-longwave-radiation-daily

EARTH'S ENERGY BALANCE

Big Picture: Equilibrium between incoming solar radiation and outgoing heat energy is critical to life on earth. If the earth loses more energy than it gains from the sun, the earth cools down. If the earth collects more energy than it loses, it heats up. The energy in the earth's atmosphere determines its state and dictates its climates. Human influence is changing this balance at a rate no natural cycle can explain. Land use affects the energy balance. Deforestation, construction, and agricultural practices change the surface of the earth, which changes the albedo and the way the earth responds to incoming solar radiation. But the biggest factor influencing climate change appears to be the retention of heat energy because of greenhouse gases.

FACILITATING THE ACTIVITY

Question: What percentage of the incoming solar energy becomes part of the earth's energy system?

ANSWER:
71% enters earth's atmosphere.

Question: What experiences the most solar heating: the atmosphere or earth's surfaces?

ANSWER:
Surfaces

Question: What experiences the most radiative cooling: the atmosphere or earth's surfaces?

ANSWER:
Atmosphere

Question: How much radiation (percent heat equivalent) must be radiated back into space for earth's temperature to remain stable (over long periods of time)?

ANSWER:
71%—the same amount that enters earth's atmosphere must be radiated back to space for the planet's average temperature to remain stable.

Question: Create two pie charts to show the amount of sunlight absorbed and heat equivalent radiated from the atmosphere and surfaces of the earth. Be sure to include titles and units (% or W/m2).

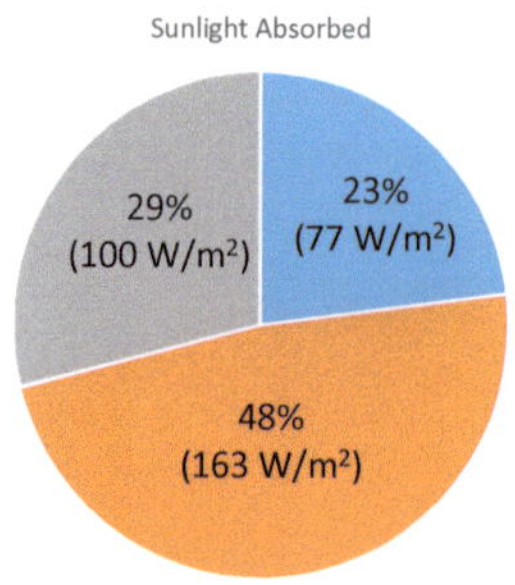

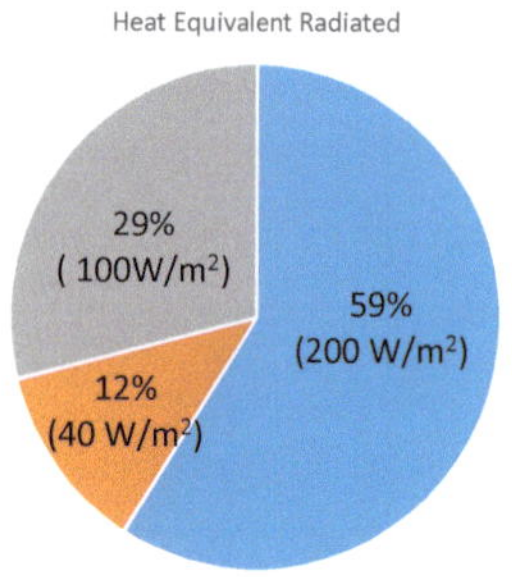

Sunlight Absorbed

Measurement	%	W/m2
Atmosphere	23	77
Surfaces	48	163
Reflected	29	
	100	

Heat Equivalent Radiated

%	W/m2
59	200
12	40
29	
100	

Question: Draw a bar graph that shows the albedo of the following: Fresh snow, dark soil, charcoal, landforms (high value), ocean surface, earth (general), landforms (low value), and forests. Show the albedos from highest to lowest.

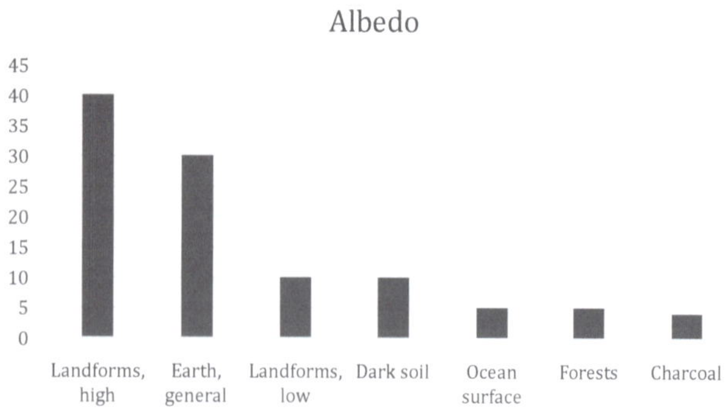

Question: Describe two results of solar radiation that is absorbed on earth's surfaces.

ANSWER:

Solar radiation warms surfaces, causes evaporation and evapotranspiration, and creates water vapor.

Question: How might the following activities affect the earth's energy balance?

ANSWERS:

Cutting down forests: Changes in the surface could change how sunlight is absorbed. Forests use CO_2 so more stays in the air, which could warm the atmosphere.

Using energy sources like wind, water, and solar energy: This reduces the need to burn fossil fuels so reduces greenhouse gases and cools the atmosphere.

Producing more efficient cars and trucks: This reduces the need to burn fossil fuels so reduces greenhouse gases and cools the atmosphere.

Burning more fossil fuels: This increases the greenhouse gases warming the atmosphere.

Constructing more buildings, roads, and parking lots: These absorb more energy, warming the atmosphere.

Planting more trees and creating more green space: More plants use more CO_2, which reduces greenhouse gases, cooling the atmosphere.

Recycling: Recycling reduces the use of energy, cooling the atmosphere.

Question: What other activities or concerns have you heard about or researched that might affect earth's climate?

ANSWER:

Open ended—no right or wrong answers.

SHARE WHAT HAPPENED

Question: Describe the earth's energy balance.

ANSWER:

Energy comes in from the sun (solar radiation), is reflected and absorbed (atmospheric or surface), and leaves the earth as longwave radiation.

APPLY

Question: How does the energy balance help explain climate change?

ANSWER:

The earth's climate remains stable as long as incoming solar radiation equals the heat equivalent radiation leaving earth. But because greenhouse gases retain heat, average global temperatures are slowly rising.

GENERALIZE TO YOUR LIFE

Question: Why is it important for people to understand things that affect the earth's energy balance?

ANSWER:
Humans have increased greenhouse gases to the point they are affecting our climate. It is up to people to decide if they are going to continue to add to the problem or to solve it. However, individual actions will not be enough. Governments have to enact policies to ensure that everyone—individuals and companies—works to solve this problem. (Search online for "tragedy of the commons" for more information on this and other environmental issues that need everyone to help solve.)

LIFE SKILLS

- Learning to learn; wise use of resources
- Weather Skills: The sun provides the energy that drives earth's climates so that there is equilibrium between energy in and energy out of the system. Human influences can change the dynamics of the atmospheric energy budget.
- Indiana Standards: ES.4.2, ES.2.3, ESA.2.3, ESA.3.3
- NGSS: ESS2.D, HS-ESS2-4, ESS.D, HS-ESS3-6
- Success indicator: Youth gain knowledge of the dynamics of the energy budget and the factors that can change the equilibrium of that system.

SUPPLEMENTAL INFORMATION AND RESOURCES

Energy Budget: https://mynasadata.larc.nasa.gov/basic-page/earths-energy-budget

IMPACTS OF CLIMATE CHANGE

Big Picture: In a 2007 report, the Intergovernmental Panel on Climate Change stated:

"Climate change is a global environmental challenge stemming from higher concentrations of so-called 'greenhouse gases' (GHG) in the atmosphere, due primarily to the burning of fossil fuels. These gases trap outgoing radiation from the earth that would normally escape into space, returning much of that energy back toward the earth's surface. This 'greenhouse effect' warms the planet and makes life on earth possible. As GHG concentrations increase in the atmosphere, however, most scientists are concerned about the higher levels of energy in the earth's climate system that could lead to higher global temperatures, altered weather patterns, changes in water salinity and sea level, and other disruptive changes in the ecological and climatological systems that support human and non-human life." (IPCC, 2007)

Scientists are researching climate change in general and the specific environmental impact that might be expected. These studies predict changes and future problems if the earth's average temperatures continues to rise. In this activity, youth do their own research to investigate the effect of climate change on one plant or animal species and other factors that might impact the environment.

FACILITATING THE ACTIVITY

No references are provided. 4-H members should do their own searches for data and information. Youth should be encouraged to find their information in at least three reputable, peer-reviewed sources from university or governmental scholars.

SHARE WHAT HAPPENED

ANSWER:
Questions are interpretive—for discussion only.

LIFE SKILLS

- Learning to learn; goal setting; planning/organizing
- Weather Skills: Research an environmental impact of climate change.
- Indiana Standards: B.4.2
- NGSS: ESS3.D, HS-ESS2-2, HS-ESS3-5, HS-PS3.B

- Success Indicator: Youth present data and research supporting ideas of environmental impacts of climate change.

REFERENCE

IPCC, 2007: Climate Change 2007: Synthesis Report. Contribution of Working Groups I, II and III to the Fourth Assessment Report of the Intergovernmental Panel on Climate Change [Core Writing Team, Pachauri, R.K and Reisinger, A. (eds.)]. IPCC, Geneva, Switzerland, 104 pp. Available online at the IPCC Synthesis Report website: www.ipcc.ch/site/assets/uploads/2018/03/ar4_wg2_full_report.pdf.

INVESTIGATING CLIMATE CHANGE

Big Picture: Studying climate and climate change can be challenging. Observing and tracking climate change can only occur over a period of years. A lot of data has been recorded over the last 100 years. This data is continually being updated. Governmental websites share this data, usually in graphs and charts that people can be use to visualize trends. Youth are asked to explore the major governmental agencies that collect weather data—NOAA, EPA, NASA—and create their own chart or graphic that shows a 30-year record of measurement of two types of weather data.

FACILITATING THE ACTIVITY

Encourage youth to consider data that will center discussion on the causes and effects of climate change; for example, more greenhouse gases have increased global temperature, and as a result, sea levels are rising. This could lead to several other effects.

SHARE WHAT HAPPENED

ANSWER:
Questions are interpretive, for discussion only.

APPLY

Question: How do you think your results would look if you were a 4-H member in 1970 and collected the previous 30 years of weather data?

ANSWER:
A discussion of climate change 30–40 years ago might be much different than that same discussion today. Collecting data each year leads to new and more insightful conclusions. In this activity websites and examples are provided. These websites might change, but the information is still there in some form. Use the information to show cause-and-effect relationships among data sets.

GENERALIZE TO YOUR LIFE

Question: Why is it important that others understand weather trends?

ANSWER:
Weather trends determine climate. It is important that everyone understands how climate change affects all plants and animals, because everyone needs to be part of the solution. Having 25% of people and industry change will not solve our problems, if others do not. This is called the tragedy of the commons.

LIFE SKILLS

- Critical thinking; planning/organizing/wise use of resources
- Weather Skills: Use climate data to study how weather data is used to determine climate.
- Indiana standards: ES.3.3, ES.4.6, ESA.3.3ESA.1.2, ESA.1.5
- NGSS: ESS2.A, HS-ESS2-2, HS-ESS2-7, ESS3.D, HS-ESS3-5
- Success Indicator: Youth use data sources to discuss relationships in the data.

THE SUN-EARTH RELATIONSHIP

Big Picture: Climate change skeptics point to natural orbital changes and Volstok ice core data to dismiss human-caused global warming. The Volstok ice core data is a recording of the temperature and atmospheric gas composition of the last 400,000 years. In this activity youth are given the data. By studying it, they can see a repeated, slow progression toward a cooler earth with intervals of rapid warming. Carbon dioxide data directly corresponds with the temperature data, indicating that the relationship of carbon dioxide and temperature is not a coincidence.

In the 1920s Serbian scientist Milutin Milankovitch developed a theory to explain possible climate changes caused by the earth's changing patterns of orbit and tilt in relationship to the sun. He was especially interested in explaining the occurrence of ice ages. His theory, the Milankovitch cycles, was mostly ignored until the Volstok data was presented in the 1970s. Since then climate and atmospheric scientists have taken a serious look at how the Milankovitch cycles could explain the Volstok data. Initially, the change in CO_2 levels led them to think that those levels cause global warming and cooling, but that is not the case. Orbital cycles changing the amount of radiation from the sun causes the warming and cooling. As the earth cools, organic material freezes in the soils and deep oceans. This slow process reduces the atmosphere's ability to retain heat through the greenhouse effect, causing further cooling. When orbital cycles bring increased radiation to the earth, frozen stores of organic material thaw and release methane into the atmosphere. Methane is a strong greenhouse gas that degrades into CO_2, so CO_2 levels increase as a result of the warming.

Looking at the Volstok data, we see a repeated warming of about 8°C over about 20,000 years. Using our current data, we see more than a 1°C rise over the last 100 years, with an increased rate of warming over the last 30 years. This cannot be explained by the slow changes of the earth's orbit. Burning fossil fuels is a much more likely explanation of rapidly increasing CO_2 and current warming.

FACILITATING THE ACTIVITY

Question: What do you think?

ANSWER:
Youth are not expected to know these answers. These answers will vary based on youth's prior knowledge and beliefs.

From the data

Question: How many cooling periods do you observe during the past 400,000 years in Figure 68?

ANSWER:
Four

Question: Which change seems to occur at a faster rate, warming or cooling?

ANSWER:
Warming

Question: What did Milankovitch's theory try to explain?

ANSWER:
His theory tried to explain the occurrence of ice ages.

Precession

Question: Which cycle explains the direction of the earth's tilt?

ANSWER:
Precession

Question: Which cycle explains the shape of the earth's orbit?

ANSWER:
Eccentricity

Question: Which cycle explains the angle of the earth's tilt?

ANSWER:
Obliquity

Eccentricity

Question: What is the percentage of eccentricity of the earth's orbit?

ANSWER:
Earth's eccentricity = 3 million miles / 93 million miles = 0.0322 = 3%

Question: What effect would a more circular orbit have on seasons?

ANSWER:
A more circular orbit would cause less severe seasonal changes.

Question: What effect would a more eccentric orbit have on seasons?

ANSWER:
A more eccentric orbit would cause more severe seasonal changes.

Question: How long does it take for this cycle to repeat?

ANSWER:
This cycle takes about 100,000 years to repeat.

Obliquity

Question: At a maximum tilt of 24.5°, what will happen to seasons in the higher latitudes?

ANSWER:
At the maximum tilt, seasons are severe.

Question: At a minimum tilt of 22.1° what will happen to seasons in the higher latitudes?

ANSWER:
At minimum tilt, seasons are less severe.

Question: How long does it take for the earth to complete the tilt cycle?

ANSWER:
This cycle takes about 40,000 years.

Precession

Question: During which of our Northern Hemisphere seasons is earth closest to the sun?

ANSWER:
Earth is closest to the sun in the winter.

Question: How long does this cycle take to complete?

ANSWER:
This cycle takes about 26,000 years.

Question: When the earth is closest to the sun during our summer, what effect might that have on our seasons?

ANSWER:
This would cause more severe seasons.

Question: What status of the Milankovitch cycles are expected to create the next ice age?

ANSWER:
Minimum eccentricity + minimum obliquity + precession, such that earth is closest the sun during spring.

Question: Could these cycles cause changes during a period of hundreds of years or thousands of years?

ANSWER:
Thousands of years—the Milankovitch cycles follow patterns on a solar system scale.

Question: Figure 71 shows a rise in global temperatures compared to the global average since 1880. Could changes in the Milankovitch cycles explain the recent rapid global temperature rise shown in Figure 73? Why or why not?

ANSWER:
No, because the Milankovitch cycles occur over thousands of years, but the rise in the average global temperature has happened over the last 50 years.

Question: Figure 70 shows the temperature has risen about 8°C over the past 20,000 years. Figure 73 shows the temperature has risen over 1.5°C in the last 130 years. Which time period is showing the most rapid temperature rate?

ANSWER:
The last 130 years: 8/20,000 = 0.0004; 1.5/130 = 0.0115

SHARE WHAT HAPPENED

Question: What are the Milankovitch cycles?

ANSWER:
Milankovitch suggested that three cycles describe the sun-earth relationship: obliquity, eccentricity, and precession.

APPLY

Question: Why is it warmer in the summer than in the winter in the United States?

ANSWER:
The earth's tilt is responsible for the changing seasons and their differences in temperature. During the summer the Northern Hemisphere is tilted toward the sun; in the winter, it is tilted away.

GENERALIZE TO YOUR LIFE

Question: Events that follow cycles will occur again. How can genealogy describe a family cycle?

ANSWER:
Genealogy describes the family cycle of births, marriages, and deaths. Some people can trace their family tree back through many generations.

LIFE SKILLS

- Critical thinking; problem solving
- Weather Skills: Recognize how the Milankovitch cycles cause changes in the earth's orbit and orientation.
- Indiana Standards: ESA.1.2, ESA.3.3, ES.4.2, ES.4.3
- NGSS: ESS2.D, ESS3.D, HS-ESS3-6, ESS1.B
- Success Indicator: Youth recognize that cyclical conditions cannot explain the current, persistent change in global temperature increases.

SUPPLEMENTAL INFORMATION AND RESOURCES

Adapted from Activities for Conceptualizing Climate and Climate Change: The Milankovitch Cycles and Their Effect on Climate Change, http://iclimate.org/ccc/index.asp

NCDC mean annual global temperatures, https://www.ncei.noaa.gov/access/monitoring/monthly-report/global/202113

Milankovitch (Orbital) Cycles and Their Role in Earth's Climate, https://science.nasa.gov/science-research/earth-science/milankovitch-orbital-cycles-and-their-role-in-earths-climate

Milankovitch Cycles and Glaciation, https://science.nasa.gov/learn/heat/resource/milankovitch-cycles-and-their-role-in-earths-climate-interactives

NASA Explores, www.nasaexplores.com

NOAA Earth System Research Laboratory, https://esrl.noaa.gov

Question: Is there any question that surface temperatures in the United States have been rising rapidly during the last 50 years?

ANSWER:
No. Even if NOAA did not have weather-observing stations across the United States, the impacts of the warming are clear and present. For example, lake and river ice is melting earlier in the spring and forming later in the fall. Plants are blooming earlier in the spring. Mountain glaciers are melting. Coastal temperatures are rising. A multitude of species of birds, fish, mammals, and plants are extending their ranges northward and, in mountainous areas, uphill toward cooler areas.

SUNSPOT CYCLE

Big Picture: Some global-warming skeptics blame rising global temperatures on natural factors. Because the earth receives its energy from the sun, variations in solar energy output must also be investigated. In this activity, youth discover that a decrease in solar output might be to blame for the Little Ice Age in the late 17th century. They look at current data to see how the recent state of the sun would suggest that global temperatures should be declining, but instead are staying at their highest levels. The only persistent change that can explain the data is the increasing amount of greenhouse gases, especially CO_2 and methane, added to the atmosphere.

FACILITATING THE ACTIVITY

Question: Were temperatures increasing, staying the same, or decreasing from 1880 through 1920?

ANSWER:
Temperatures were increasing.

Question: Were temperatures increasing, staying the same, or decreasing from 1920 through 1977?

ANSWER:
Temperatures were staying the same.

Question: Were temperatures increasing, staying the same, or decreasing from 1977 through 2012?

ANSWER:
Temperatures were increasing.

Question: During which time period do you see the most change occurring?

ANSWER:
The most change happened from 1977 through 2012.

Question: Which five years were the warmest?

ANSWER:
Answers may vary because of the table, but they are 1998, 2002, 2005, 2007, and 2010.

Question: What effect do you think a solar maximum has on temperatures on earth?

ANSWER:
A solar maximum should cause a rise in temperature.

Question: What effect do you think a solar minimum has on temperatures on earth?

ANSWER:
A solar minimum should cause a decrease in temperature.

Question: How might sunspot activity have contributed to the Little Ice Age?

ANSWER:
The cycle was in a long-term minimum, which reduces the amount of energy reaching the earth.

Question: Was sunspot activity increasing or decreasing from 2000 to 2014?

ANSWER:
Sunspot activity was decreasing.

Question: Was global temperature increasing or decreasing from 2000 to 2010?

ANSWER:
Global temperatures were increasing.

Question: What should have happened to global temperatures between 2000 and 2010 as a result of sunspot activity?

ANSWER:
Because of sunspot activity, global temperatures should have decreased.

Question: The earth experienced three of the highest recorded temperatures in 2005, 2007, and 2010. What was the sunspot activity like during these years?

ANSWER:
Sunspot activity was low.

Question: Does sunspot activity explain the most recent increase in temperature?

ANSWER:
No, sunspot activity does not explain the recent increase in temperature.

Question: How does sunspot activity affect the amount of solar radiation given off by the sun?

ANSWER:
During times of high sunspot activity, the sun gives off more radiation.

Question: How might increased radiation from the sun (sunspot cycle) affect the earth's temperature?

ANSWER:
Increased radiation should cause the earth to warm.

Question: Can the sunspot cycle alone explain the recent increase in global temperature?

ANSWER:
No, cooling should happen when sunspots are low in number.

Question: What other factors could contribute to global temperatures?

ANSWER:
Greenhouse gases and volcanoes can contribute to global temperatures.

Question: If Mount Pinatubo hadn't erupted in 1991, what might have happened to temperatures?

ANSWER:
Temperatures might have risen even more.

Question: What is the only persistent change that could explain the increases in temperature?

ANSWER:
The only persistent change is the increase of greenhouse gases.

SHARE WHAT HAPPENED

Question: What are sunspots?

ANSWER:
Sunspots are areas on the sun that are cooler than the areas around them. They occur when the magnetic field of the sun becomes active.

APPLY

Question: What happens to the earth when there is little sunspot activity?

ANSWER:
When sunspot activity is low, the earth becomes unusually cold, as in the Little Ice Age, 1645–1715.

GENERALIZE TO YOUR LIFE

Question: Is an increased number of sunspots (solar max) responsible for climate change?

ANSWER:
No, as shown by comparing Figures 74, 75, and 76.

LIFE SKILLS

- Critical thinking; problem solving
- Weather Skills: Identify the relationship of sunspot cycles to sun radiation output and the resulting effect on global atmospheric temperatures.
- Indiana Standards: ESA.1.2, ESA.3.3, ES.4.2

  NGSS: ESS2.D, ESS3.D, HS-ESS3-6
- Success Indicator: Youth recognize that changes in solar activity cannot explain persistent increases in global temperature.

SUPPLEMENTAL INFORMATION AND RESOURCES

Adapted from Activities for Conceptualizing Climate and Climate Change, http://iclimate.org/ccc/index.asp

The Sunspot Cycle, http://solarscience.msfc.nasa.gov/SunspotCycle.shtml

NCDC mean annual global temperatures, https://www.ncei.noaa.gov/access/monitoring/monthly-report/global/202113

NASA Earth Observatory, https://science.nasa.gov/earth/earth-observatory/

NASA Explores, www.nasaexplores.com

Sunspot image, https://science.nasa.gov/sun/sunspots

NOAA Earth System Research Laboratory, https://www.esrl.noaa.gov

VOLCANOES AND GLOBAL WARMING

Big Picture: Youth investigate the type and amount of materials ejected in a volcanic eruption and how these materials might affect the atmosphere. Based on data and information, youth compare the role of volcanic activity in changing global temperatures to the effect of human activity. Some people think volcanoes contribute more carbon dioxide than human activity contributes. In reality, volcanoes contribute an amount equal to a little more than 1% of the carbon dioxide from human actions. In this activity youth use data to discover this, and to see that volcanoes actually have a cooling effect on global temperatures.

FACILITATING THE ACTIVITY

Question: Some youth might need help calculating the scale to compare volcanic CO_2 emissions to human activity CO_2 emissions.

ANSWERS:
The open-ended What do you think? questions are intended to have youth think about these issues before beginning the activity. Answers will be based on the youth's background knowledge and beliefs.

Explore and explain

Question: Which greenhouse gases are added to the atmosphere during a volcanic eruption?

ANSWER:
Carbon dioxide and water are added during a volcanic eruption.

Question: How do sulfate aerosols affect the temperature of the atmosphere?

ANSWER:
Sulfate aerosols reflect sunlight back into space, reducing the temperature of the atmosphere.

Question: How does volcanic ash affect the temperature of the atmosphere?

ANSWER:
Volcanic ash reflects sunlight back into space, reducing the temperature of the atmosphere.

Look at the data

Question: Calculate the average for each gas and enter the value in column 5 of Table 1.

ANSWERS:
H_2O—70.5%
CO_2—20.5%
SO_2—6.9%

Question: On average, which gas—H_2O, CO_2, or SO_2—composes the highest percentage of the ejected gases in the volcanoes listed in Table 1?

ANSWER:
These volcanoes ejected H_2O, water vapor, the most on average.

Question: Which volcano ejected the greatest percentage of carbon dioxide?

ANSWER:
Kilauea ejected the most carbon dioxide (48.9%).

Question: Which volcano ejected the least amount of carbon dioxide?

ANSWER:
Momotombo ejected the least amount of carbon dioxide (1.44%).

Question: Why might there be so much variation in the composition of gases that come from volcanic eruptions?

ANSWER:
Interpretive—no right or wrong answers.

Question: How much more CO_2 do humans create than volcanoes (percentage)?

ANSWER:
Humans: 3.0 x 1010 tons

Volcanoes: 2.0 x 108 tons

Human activity / volcanic activity

= 3.0 x 1010 tons / 1.85 x 108 tons

= 3 x 102 / 1.85

= 300/1.85

= 162.2 times or 1622% more CO_2 created by humans than volcanoes!

Question: How much CO_2 do volcanoes create compared to humans (percentage)?

ANSWER:
Humans: 3.0 x 1010 tons

Volcanoes: 2.0 x 108 tons

Volcanic activity / human activity

= 1.85 x 108 tons / 3.0 x 1010 tons

= 1.85 / 3 x 102

= 1.85 / 300

= 0.0062 or 0.62 % of the CO_2 created by humans!

Question: Compared to human activity, how important are volcanoes in adding carbon dioxide to the atmosphere?

ANSWER:
Volcanoes add very little carbon dioxide compared to human activity.

Question: What is the effect on global temperatures of the sulfate aerosols formed by volcanoes?

ANSWER:
Sulfate aerosols have a cooling effect on global temperatures.

Question: Using the ideas of greenhouse gases, volcanic ash, and sulfate aerosols, summarize the effect of volcanoes on global temperatures.

ANSWER:
Volcanoes contribute carbon dioxide to the atmosphere, but a rate much smaller than human activity. The volcanic ash and sulfate aerosols ejected into the atmosphere have a cooling effect on the global temperatures by reflecting sunlight. Sulfate aerosols are ejected into the upper atmosphere, where they might persist for years.

SHARE WHAT HAPPENED

Question: Do volcanoes cause global warming?

ANSWER:
No

Question: Can they cause global cooling?

ANSWER:
Yes

APPLY

Question: How does an understanding of chemistry help understand climate change?

ANSWER:
Chemistry is important to understanding climate change—how water vapor combines with other gases to create hydrochloric acid and hydrofluoric acid, and the effects of these and other chemicals such as sulfur dioxide.

GENERALIZE TO YOUR LIFE

Question: Why do some people make arguments that are not based on data?

ANSWER:
Interpretive question. Katherine Hayhoe, a scientist, and her husband Andrew Farley, a Baptist minister, address some of these arguments in *A Climate for Change: Global Warming Facts for Faith-Based Decisions*.

LIFE SKILLS

- Critical thinking; problem solving
- Weather Skills: Youth recognize the effect of volcanic eruptions on global temperatures.
- Indiana Standards: ESA.1.2, ESA.3.3, ES.4.2
- NGSS: ESS2.D, ESS3.C, ESS3.D, HS-ESS2-2, HS-ESS2-4
- Success Indicator: Youth understand that sulfate aerosols have a lasting cooling effect on global temperatures, and that the amount of carbon dioxide ejected by volcanoes is minor compared to carbon dioxide contributed by human activity.

SUPPLEMENTAL INFORMATION AND RESOURCES

Adapted from Activities for Conceptualizing Climate and Climate Change, http://iclimate.org/ccc/index.asp

USGS Volcanic Hazards, www.usgs.gov/programs/VHP

Glossary: USGS Volcanoes and the Weather, https://www.usgs.gov/glossary/volcano-hazards-program-glossary

NEXT GENERATION SCIENCE STANDARDS

Weather and Climate Science Education for Grades 3-12 primarily addresses the Next Generation Science Standard (NGSS) educational disciplines: Earth and Space Science and the Core Ideas:

ESS1 Earth's Place in the Universe
ESS2 Earth's Systems
ESS3 Earth and Human Activity

The discipline of Physical Science (PS) and the Core Ideas below are touched on in a few activities.

PS1 Matter and Its Interactions
PS3 Energy

The table below lists Sub Ideas that each activity touches upon and performance expectation codes, when addressed. See NGSS Text for a description of the performance expectations.

ACTIVITY	SUB IDEA	PERFORMANCE EXPECTATION CODE
LEVEL 1		
Comparing Climates	ESS2.D	3-ESS2-1 3-ESS2-2
Country of Colors	ESS2.D	MS-ESS3-5
Defining Weather Words	ESS2.D	
Earth's Surfaces	ESS2.D	5-ESS2-1 MS-ESS2-6
H2O	ESS2.D	MS-ESS2-5
Invisible Air	ESS2.D PS1.A PS3.A	5-PS1-1 MS-ESS2-5 MS-PS3-4
Reading About Wild Weather	ESS2.D	
'Tis the Season	ESS2.D	3-ESS2-1
Watching the Wind	ESS2.D	3-ESS2-1
Weather Affects Plans	ESS2.D	5-ESS3-1
Weather Alerts	ESS2.D	5-ESS3-1
Weather or Climate?	ESS2.D	3-ESS2-1
Where Is the Heat?	ESS1.B	5-ESS1-2

ACTIVITY	SUB IDEA	PERFORMANCE EXPECTATION CODE
LEVEL 2		
Air Pressure	ESS2.D	5-PS1-1
Carbon Footprints	ESS2.D ESS3.D	MS-ESS2-5 MS-ESS3-5
Cloud Formation	ESS2.D	5-ESS2-1 5-PS1-1
Cloud Types	ESS2.D	MS-ESS2-5
Cold Fronts	ESS2.D	MS-ESS2-5
Earth's Rotation	ESS2.D	
Global Winds	ESS2.D	MS-ESS2-6
The Greenhouse Effect	ESS2.D ESS3.D	MS-ESS2-5 MS-ESS3-5
Humidity	ESS2.C ESS2.D PS1.A	MS-ESS2-5 MS-PS3-4
Hurricanes	ESS2.D	MS-ESS2-5 MS-ESS2-6
Making Weather Instruments	ESS2.D	3-ESS2-1 MS-ESS2-5
Mini-Tornado	PS1.A	MS-PS3-4
Out of the Dust	ESS2.D	MS-ESS2-5
Seasons	ESS1.B	5-ESS1-2 MS-ESS2-5 MS-ESS2-6
Using Weather Instruments to Collect Data	ESS2.D	3-ESS2-1 MS-ESS2-5

ACTIVITY	SUB IDEA	PERFORMANCE EXPECTATION CODE
LEVEL 3		
WEATHER		
Air Masses and Front	ESS2.D	5-ESS2-1 MS-ESS2-5
Isaac's Storm	ESS2.D ESS3.B	
Monitoring Weather	EES2.D	5-ESS1-2 MS-ESS2-5
Pressure Systems	ESS2.D	MS-ESS2-5
Weather Station Models	ESS2.D	MS-ESS2-5
Weather in the Troposphere	ESS2.D	HS-ESS2-4
Windchill and Heat Index	ESS2.D	MS-ESS2-5
CLIMATE		
Climate and Climographs	ESS2.D ESS3.D	3-ESS2-2
Drought Monitoring	ESS2.D	MS-ESS2-5 MS-ESS2-6
Energy in the Atmosphere	ESS2.D ESS3.D	MS-ESS2-6
Energy Balance	ESS2.D ESS3.D	HS-ESS2-4 HS-ESS3-6
Impacts of Climate Change	HS-PS3.B ESS3.D	HS-ESS2-2 HS-ESS3-5
Investigating Climate Change HS-ESS2-2 HS-ESS3-5		ESS2.A ESS3.D
		HS-ESS2-7
The Sun–Earth Relationship	ESS1.B ESS2.D ESS3.D	HS-ESS3-6
Sunspot Cycle	ESS2.D ESS3.D	HS-ESS3-6
Volcanoes	ESS3.C ESS2.D ESS3.D	HS-ESS2-2 HS-ESS2-4

Reference: Next Generation Science Standards for States, By States, Volume 1, The Standards (ISBN 13: 978-0-309-27227-8; ISBN 10: 0-309-27227-0; Library of Congress Control Number: 2013939525)

SCIENCE STANDARDS

PERFORMANCE STANDARDS, LISTED BY DISCIPLINE, CORE IDEA

Earth's Place in the Universe, Core Idea: ESS1

Sub Idea: ESS1.B Earth and the Solar System

Performance Expectation code 5-ESS1-2

Represent data in graphical displays to reveal patterns of daily changes in the length and direction of shadows day and night and seasonal appearance of some stars in the night sky. (The orbits of Earth around the sun and of the moon around Earth, together with the rotation of Earth about an axis between its north and south poles, cause observable patterns. These include day and night; daily changes in the length and direction of shadows; and different positions of the sun, moon, and stars at different tines of the day, month, and year.)

- Level 1, Where Is the Heat?
- Level 2, Seasons
- Level 3, Monitoring Weather

Earth's Systems, Core Idea: ESS2

Sub Idea: ESS2.C—The Roles of Water in Earth's Surface Processes

Sub Idea: ESS2.D—Weather and Climate

Performance Expectation code: 3-ESS2-1

Represent data in tables and graphical displays to describe typical weather conditions expected during a particular season. (Scientists record patterns of the weather across different times and areas so that they can make predictions about what kind of weather might happen next.)

- Level 1, Comparing Climates
- Level 1, 'Tis the Season
- Level 1, Watching the Wind
- Level 1, Weather or Climate?
- Level 2, Making Weather Instruments
- Level 2, Using Weather Instruments to Collect Data

Performance Expectation code: 3-ESS2-2

Obtain and combine information to describe climates in different regions of the world. (Climate describes a range of an area's typical weather conditions and the extent to which those conditions vary over years.)

- Level 1, Comparing Climates
- Level 3, Climate and Climographs

Performance Expectation code: 5-ESS2-1

Develop a model using an example to describe ways in which the geosphere, biosphere, hydrosphere, and/or atmosphere interact. Winds and clouds in the atmosphere interact with the landforms to determine patterns of weather.)

- Level 1, Earth's Surfaces
- Level 2, Cloud Formation
- Level 3, Air Masses and Fronts

Performance Expectation code: MS-ESS2-5

Collect data to provide evidence of how motions and complex interactions of air masses result in changes in weather conditions. [Emphasis is on how air masses flow from regions of high pressure to low pressure, causing weather at a fixed location to change over time and how sudden changes in weather can result when different air masses collide. Examples of data can be provided to students (such as weather maps, diagrams, and visualizations) or obtained through laboratory experiments (such as with condensation). Assessment does not include recalling the names of cloud types or weather symbols.

- Level 1, H2O
- Level 1, Invisible Air
- Level 2, Carbon Footprints
- Level 2, Cloud Types
- Level 2, Cold Fronts
- Level 2, The Greenhouse Effect
- Level 2, Humidity
- Level 2, Hurricanes
- Level 2, Making Weather Instruments
- Level 2, Out of the Dust
- Level 2, Seasons
- Level 2, Using Weather Instruments to Collect Data
- Level 3, Air Masses and Fronts
- Level 3, Monitoring Weather
- Level 3, Pressure Systems
- Level 3, Weather Station Models
- Level 3, Windchill and Heat Index
- Level 3, Drought Monitoring

Performance Expectation code: MS-ESS2-6

Develop and use a model to describe how unequal heating and rotation of the Earth cause patterns of atmospheric and oceanic circulation that determine regional climates. [Emphasis of atmospheric circulation is on the sunlight-driven latitudinal banding, the Coriolis effect, and resulting prevailing winds; emphasis of ocean circulation is on the transfer of heat by the global ocean convection cycle, which is constrained by the Coriolis effect and the outlines of continents. (Weather and climate are influenced by interactions involving sunlight, the ocean, the atmosphere, ice, landforms, and living things. These interactions vary with latitude, altitude, and local and regional geography, all of which can affect oceanic and atmospheric flow patterns. The ocean exerts a major influence on weather and climate by absorbing energy from the sun, releasing it over time, and globally redistributing it through ocean currents.)]

- Level 1, Earth's Surfaces
- Level 2, Global Winds
- Level 2, Hurricanes
- Level 2, Seasons
- Level 3, Drought Monitoring
- Level 3, Energy in the Atmosphere

Performance Expectation code: HS-EES2-2

Analyze geoscience data to make the claim that one change to Earth's surface can create feedbacks that cause changes to other Earth systems.

- Level 3, Impacts of Climate Change
- Level 3, Investigating Climate Change
- Level 3, Volcanoes

Performance Expectation code: HS-EES2-4

Use a model to describe how variations in the flow of energy into and out of Earth's systems result in changes in climate. (Examples of the causes of climate change differ by timescale, changes in human activity, ocean circulation, solar output, changes to Earth's orbit and the orientation of its axis.)

- Level 3, Weather in the Troposphere
- Level 3, Energy Balance
- Level 3, Volcanoes

Performance Expectation code: HS-ESS2-7

Construct an argument based on evidence about the simultaneous co-evolution of Earth's systems and life on Earth. (Emphasis is on the dynamic causes, effects, and feedbacks between the biosphere and Earth's other systems.)

- Level 3, Investigating Climate Change

Earth and Human Activity, Core Idea: ESS3

Sub Idea: ESS3.C Human Impacts on Earth Systems

Sub Idea: ESS3.D Global Climate Change

Performance Expectation code: 5-ESS3-1

Obtain and combine information about ways individual communities use science ideas to protect the Earth's resources and environment. (Individuals and communities are doing things to help protect Earth's resources and environments.)

- Level 1, Weather Affects Plans
- Level 1, Weather Alerts

Performance Expectation code: MS-ESS3-5

Ask questions to clarify evidence of the factors that have caused the rise in global temperatures over the past century. [Examples of factors include human activities (such as fossil fuel combustion, cement production, and agricultural activity) and natural processes (such as changes in incoming solar radiation or volcanic activity). Examples of evidence can include tables, graphs, and maps of global and regional temperatures, atmospheric levels of gases such as carbon dioxide and methane, and the rates of human activities. Emphasis is on the major role that human activities play in causing the rise in global temperatures.]

- Level 1, Country of Colors
- Level 2, Carbon Footprints
- Level 2, The Greenhouse Effect

Performance Expectation code: HS-ESS3-5

Analyze geoscience data and the results from global climate models to make an evidence-based forecast of the current rate of global or regional climate change and associated future impacts to Earth's systems. [Examples of evidence are for climate changes (such as precipitation and temperature) and their associated impacts.]

- Level 3, Impacts of Climate Change
- Level 3, Investigating Climate Change

Performance Expectation code: HS-EES3-6

Use a computational representation to illustrate the relationships among Earth systems and how those relationships are being modified due to human activity. (An example of the far-reaching impacts from a human activity is how an increase in atmospheric carbon dioxide results in an increase in photosynthetic biomass on land and an increase in ocean acidification with resulting impacts on sea organism health and marine populations. Assessment does not include running computational representations, but is limited to using the published results of scientific computational models.)

- Level 3, Energy Balance
- Level 3, The Sun-Earth Relationship
- Level 3, Sunspot Cycle

DISCIPLINE: PHYSICAL SCIENCES (PS)

PS1 Matter and Its Interactions, Core Idea: PS1

Sub Idea: PS1.A Structure and Properties of Matter

Performance Expectation code: 5-PS1-1

Develop a model to describe that matter is made of particles too small to be seen. (Examples of evidence could include adding air to expand a basketball, compressing air in a syringe, dissolving sugar in water, and evaporating salt water. Matter of any type can be subdivided into particles that are too small to see, but even then the matter still exists and can be detected by other means. A model that shows gasses are made from matter particles that are too small to see an that are moving freely around in space can explain many observations.)

- Level 1, Invisible Air
- Level 1, Air Pressure
- Level 2, Cloud Formation

Energy, Core Idea: PS3

Sub Idea: PS1.A Structure and Properties of Matter

Performance Expectation code: MS-PS3-4

Plan an investigation to determine the relationships among the energy transferred, the type of matter, the mass, and the change in the average kinetic energy of the particles as measured by the temperature of the sample. (Examples of experiments could include comparing final water temperatures after different masses of ice have melted in the same volume of water with the same initial temperature.)

- Level 1, Invisible Air
- Level 2, Humidity
- Level 2, Mini-Tornado

NOTES

NOTES

NOTES

About the Authors

Natalie J. Carroll is a Purdue University professor with extensive experience in curriculum development. She and her creative writing teams have composed thirty-two natural resource, activity-based curricula for youth and twenty-three grant-funded projects. She and her co-authors have received ten national awards for their work. She has also published numerous journal articles and educational works for youth and adults, and has presented her work to hundreds of conferences in Indiana and around the world. Carroll has received seven awards for curriculum development from professional associations in addition to Purdue's Christian J. Foster Award for Contributions to K–12 STEM Education and the Frederick L. Hovde Award of Excellence in Educational Service to the Rural People of Indiana from the Indiana Farm Bureau.

Theodore Leuenberger regularly works with Purdue University faculty on a variety of publications and projects. He taught junior and senior high school biology, general science, and earth science for thirty-seven years, and contributed to seven publications, most recently "Field Studies as a Pedagogical Approach to Inquiry" in Science as Inquiry in the Secondary Setting. He has copresented his work at dozens of national and state science teacher association meetings, and he received the Purdue University Department of Education Distinguished Alumni Award in 2002.

Gretchen Leuenberger has enjoyed diverse experiences as an elementary teacher, librarian, reading specialist, and elementary-school principal. She served for forty-two years as a career curriculum specialist, presenting curriculum materials and programs at schools throughout Indiana.

www.ingramcontent.com/pod-product-compliance
Ingram Content Group UK Ltd.
Pitfield, Milton Keynes, MK11 3LW, UK
UKHW060827210526
471334UK00002B/3
9781626714052